ERGONOMICS – A SYSTEMS APPROACH

Edited by **Isabel L. Nunes**

Ergonomics – A Systems Approach
Edited by Isabel L. Nunes

Published by InTech
Janeza Trdine 9, 51000 Rijeka, Croatia

Publishing Process Manager Martina Blecic

Technical Editor Teodora Smiljanic

Cover Designer InTech Design Team

Additional hard copies can be obtained from orders@intechopen.com

Ergonomics – A Systems Approach, Edited by Isabel L. Nunes
p. cm.
ISBN 978-953-51-0601-2

For EU product safety concerns:
IN TECH d.o.o., Prolaz Marije Krucifikse Kozulić 3, 51000 Rijeka, Croatia,
info@intechopen.com or visit our website at intechopen.com.

Meet the editor

Professor Isabel L. Nunes is currently a Professor at the Faculdade de Ciências e Tecnologia – Universidade Nova de Lisboa, Portugal. She obtained her PhD in Industrial Engineering - Ergonomics, from the Universidade Nova de Lisboa. She is a member of the Computational Intelligence Research Group integrated in the Centre of Technologies and Systems – UNINOVA. Her main research interests include Ergonomics, Occupational Safety and Health and Production Systems, namely the development of artificial intelligence models using fuzzy set theory. She has published several scientific papers and book chapters. She is a patent owner, and has been a member of several scientific conference committees. She has coordinated and participated in several research projects both national and international. At her home university she teaches Ergonomics and Occupational Safety and Health and supervises PhD and MSc research work in these domains.

Contents

Preface

Ergonomics, also known as Human Factors, is a recent scientific discipline, curiously with a well-defined and official date and place of birth, July 12, 1949, in England. However the term Ergonomics, was proposed in 1857, by the Polish philosopher and naturalist Wojciech Jastrzebowski and fell into oblivion for nearly a century. The word Ergonomics results from joining the Greek words *ergon* meaning "work" and *nomos* meaning "natural laws", and conveys the concern of understanding the relationships between humans and their work environment.

Ergonomics in spite of its short existence, gathers a broad body of knowledge from different disciplines in order to fit the workplace conditions and job demands to the capabilities of workers. Its main goal is to ensure humans well-being, health, and safety while maximizing the performance of production systems. Production systems, however, are a complex combination of physical, organisational and psychosocial dimensions. Therefore Ergonomics offers principles and methods to analyse and improve this multitude of interactions. Despite production systems being the main focus of Ergonomics, its intervention extends beyond work systems, to other aspects of our daily lives, like product design, leisure or sport.

This book is the result of an InTech initiative to bring together reputable researchers from different countries that could provide an interesting and up-to-date overview of different Ergonomic research applications, practices and methodologies. The 10 chapters present the research work of 19 authors from 6 different countries, and they cover the following themes: work related musculoskeletal disorders; methods in Ergonomics; usability and user-experience design; efficiency in operations, and critical thinking skills. A brief outline of the volume is presented hereafter.

Chapter 1 offers an overview on work-related musculoskeletal disorders (WMSD). The recognition that the work may adversely affect health is not new, since musculoskeletal disorders have been diagnosed for many years in the medical field. WMSD are related with repetitive and demanding working conditions and continue to represent one of the biggest problems in industrialized countries. WMSD are a group of inflammatory and degenerative diseases of the locomotion system, which result from occupational risk factors such as repetition, force or awkward posture as well as individual and psychosocial risk factors. This chapter presents WMSD causes, pathophysiological mechanisms, characterization of the principal disorders and procedures for workplace analysis and design.

Chapter 2 presents the risk factors that contribute to musculoskeletal disorders in shoulders resulting from intensive use of computers. The risk factors of musculoskeletal disorders were revealed by assessing and analyzing workplace ergonomics, worker attitudes and experiences on the use of the computer keyboard and mouse. This was followed by an experimental data collection of muscle load, muscle force and muscular fatigue from the shoulder by Surface electromyography to validate and verify the proposed mathematical model.

Chapter 3 recognises that the management of low back pain, particularly work related injuries, is very controversial and that many different treatment approaches have been tried, ranging from osteopathic manipulations to work hardening programs. This chapter addresses low back pain reviewing the anatomical, biomechanical, and physiological mechanisms that contribute to the health of the lumbar spine with particular emphasis on the intervertebral disc (IVD); considering the mechanisms that may cause pain and dysfunction in the lumbar spine; and presenting specific strategies for prevention and management of work related low back pain based on the biomechanical and physiological response of the lumbar IVD.

Chapter 4 is devoted to the comparison of software tools for occupational biomechanics and ergonomic research. It provides a survey on selected biomechanical software tools and gives a detailed analysis and a comparison of two specialized packages, 3DSSPP and JACK as well as examples of applications where one or the other may be better suited.

Chapter 5 presents a description of methodological guidelines used to prepare a new questionnaire or to adapt an existing one. Ergonomic surveys are very important tools to evaluate and identify problems in workplaces (such as industries, hospitals, and laboratories), since strategies to tackle the ergonomic issues can be derived from their results. Therefore, the surveys should be carefully prepared to obtain information in a clear and reliable way. Usually, ergonomic surveys are based on measurement instruments (questionnaires) that are applied to workers on the workplace to collect the necessary information.

Chapter 6 discusses the need for dynamic, flexible and reconfigurable assembly systems, which are able to respond adequately to changes in the characteristics and demands of the market. The chapter presents the Walking Worker Assembly Line, in which each worker utilizes various skills and functions by travelling along the manufacturing line to carry out all the required tasks. The authors argue that this flexible manpower line (or flexible assembly line) approach is one of the promising techniques for configuring effective and productive assembly systems, responding well to the challenges of the manufacturing industry.

Chapter 7 is dedicated to usability. In recent years the knowledge media support migrated from "pen and paper" to computer-based Information Systems. This evolution introduced some technological, organizational, and methodological changes affecting the demand, workload and stress over the workers, many times in a negative way. Due to this fact usability assumed an increasing importance. This chapter presents an overview of the general principles to observe when a user-centred design is adopted, provides a summary of methods and tools that are available to support the design and evaluation of products with good usability, and offers examples of guidelines and good practices that can be adopted.

Chapter 8 discusses major challenges faced by current user-centred design practices, proposes a user experience design (UXD) framework to address these challenges, and analyses three case studies to illustrate the UXD approach and formalize the UXD processes.

Chapter 9 discusses how human operators are an integral part of automated control systems and using a systematic design approach presents a new control room that is operator-focused, in order to create a safer and securer environment, contributing to efficiency in operations.

Chapter 10 is devoted to critical thinking skills for intelligence analysis, focusing on that aspect of ergonomics research that seeks to understand how people engage in cognitive work and how to develop systems and training that best support that work. A definition and a model of critical thinking are presented.

I hope this book will encourage readers, namely academic researchers and company managers interested in Ergonomics and its applications, to pursue the challenge of transforming workplaces into safer and healthier places to work while optimizing the work system performance.

I would like to thank InTech for the invitation to be an editor and to Martina Blecic, Publishing Process Manager, for organizing this book. I would also like to thank the authors and other persons who helped and encouraged me to make this book a reality.

Isabel L. Nunes, M.S., Ph.D.
Centre of Technologies and Systems,
Faculdade de Ciências e Tecnologia,
Universidade Nova de Lisboa, Caparica,
Portugal

1

Work-Related Musculoskeletal Disorders Assessment and Prevention

Isabel L. Nunes[1] and Pamela McCauley Bush[2]
[1]Centre of Technologies and Systems,
Faculdade de Ciências e Tecnologia, Universidade Nova de Lisboa,
[2]University of Central Florida,
[1]Portugal
[2]USA

1. Introduction

Work-related musculoskeletal disorders (WMSD) related with repetitive and demanding working conditions continue to represent one of the biggest problems in industrialized countries.

The World Health Organization (WHO), recognizing the impact of 'work-related' musculoskeletal diseases, has characterized WMSD s as multifactorial, indicating that a number of risk factors contribute to and exacerbate these maladies (Sauter et al., 1993). The presence of these risk factors produced increases in the occurrence of these injuries, thus making WMSD s an international health concern. These types of injuries of the soft tissues are referred to by many names, including WMSD s, repetitive strain injuries (RSI), repetitive motion injuries (RMI), and cumulative trauma disorders (CTDs) (McCauley Bush, 2011).

WMSD are diseases related and/or aggravated by work that can affect the upper limb extremities, the lower back area, and the lower limbs. WMSD can be defined by impairments of bodily structures such as muscles, joints, tendons, ligaments, nerves, bones and the localized blood circulation system, caused or aggravated primarily by work itself or by the work environment (Nunes, 2009a).

Besides the physically demanding of the jobs the ageing of the workforce are also a contribution to the widespread of WMSD , since the propensity for developing a WMSD is related more to the difference between the demands of work and the worker's physical work capacity that decreases with age (Okunribido & Wynn 2010).

Despite the variety of efforts to control WMSD, including engineering design changes, organizational modifications or working training programs, these set of disorders account for a huge amount of human suffering due to worker impairment, often leading to permanent, partial or total disability.

WMSD have also heavy economic costs to companies and to healthcare systems. The costs are due to loss of productivity, training of new workers and compensation costs. These costs are felt globally, particularly as organizations begin to develop international partnerships for manufacturing and service roles.

Conclusions derived from the 4th European Working Conditions Survey (conducted in 2005 in 31 countries: EU27 plus Norway, Croatia, Turkey and Switzerland by European Foundation for the Improvement of Living and Working Conditions) state that about 60 million workers reportedly suffer from WMSD in Europe. Therefore, within the EU, backache seems to be the most prevalent work-related health problem, followed by overall fatigue (22.5%) and stress (22.3%). Variability among Member States' self reported backache levels are high, ranging from a maximum of 47%, in Greece, to a minimum of 10.8%, in the United Kingdom. Self-reported WMSD from the newer Member States tend to be higher: overall fatigue (40.7%) and backache (38.9%) (EUROFOUND, 2007).

The same European Foundation according to data from the 5th European Working Conditions Survey, which have collected data during 2010 from around 44,000 workers in 34 European countries (EU27, Norway, Croatia, the former Yugoslav Republic of Macedonia, Turkey, Albania, Montenegro and Kosovo) concluded that European workers remain exposed to physical hazards, which means that many Europeans' jobs still involve physical labour. For instance, 33% of workers carry heavy loads at least a quarter of their working time, while 23% are exposed to vibrations. About half of all workers (46%) work in tiring or painful positions at least a quarter of the time. Also repetitive hand or arm movements are performed by more Europeans than 10 years ago. Women and men are exposed to different physical hazards, due to gender segregation that occurs in many sectors (EUROFOUND, 2010). This report reveals also that, 33% of men, but only 10% of women, are regularly exposed to vibrations, while 42% of men, but 24% of women, carry heavy loads. In contrast, 13% of women, but only 5% of men, lift or move people as part of their work. However, similar proportions of men and women work in tiring positions (48% and 45% respectively), or make repetitive hand and arm movements (64% and 63% respectively).

WMSD are the most common occupational illness in the European Union; however, it would appear that musculoskeletal disorders directly linked to strenuous working conditions are on the decline, while those related to stress and work overload are increasing (EUROFOUND, 2010). Pain in the lower limbs may be as important as pain in the upper limbs, but there is limited research to support pain as a symptom, associated risk factors and broad evidence that has been recognized as specific lower extremity WMSD risk factors (EU-OSHA, 2010).

2. Work related musculoskeletal disorders

The recognition that the work may adversely affect health is not new. Musculoskeletal disorders have been diagnosed for many years in the medical field. In the eighteenth century the Italian physician Bernardino Ramazzini, was the first to recognize the relationship between work and certain disorders of the musculoskeletal system due to the performance of sudden and irregular movements and the adoption of awkward postures (Putz-Anderson, 1988). In old medical records is also possible to find references to a variety of injuries related to the execution of certain work. In the nineteenth century, Raynauld's phenomenon, also called dead finger or jackhammer disease, was found to be caused by a lack of blood supply and related to repetitive motions. In 1893, Gray gave explanations of inflammations of the extensor tendons of the thumb in their sheaths after performing extreme exercises. Long before the Workers' Compensation Act was passed in Great Britain (1906) and CTDs were recognized by the medical community as an insurable diagnosis, workers were victims of the trade they pursued. Since these injuries only manifest themselves after a long period of time, they often went unrecognized (McCauley Bush, 2011).

Some disorders were identified by names related with the professions where they mainly occurred, for instance 'carpenter's elbow', 'seamstress', 'wrist' or 'bricklayer's shoulder', 'washer woman's sprain,' 'gamekeeper's thumb,' 'drummer's palsy,' 'pipe fitter's thumb,' 'reedmaker's elbow,' 'pizza cutter's palsy,' and 'flute player's hand' (Putz-Anderson, 1988) (Mandel, 2003). During and after the 1960s, physiological and biomechanical strains of human tissue, particularly of the tendons and their sheaths, revealed that they were indeed associated to repetitive tasks. As a result, several recommendations have been developed for the design and arrangement of workstations, as well as the use of tools and equipment to ultimately alleviate or reduce WMSDs (McCauley Bush, 2011).

In international literature there is variability in the terminology related to WMSD. Table 1 presents some of the terms found in literature (in English) and, when identified, the countries where such designation is used. Of thing to be noted is that several of these designations are intended to translate the relationship between the disorder and the suspected causal factor or mechanism of injury.

Also the classification of the conditions allows the scientific community to understand how to treat the conditions, as well as provides information that engineers can utilize to design processes and equipment to mitigate the risk factors (McCauley Bush, 2011).

Designation	Country
Cervicobrachial Syndrome	Japan, Sweden
Cumulative Trauma Disorder	USA
Occupational Cervicobrachial Disorder	Japan, Sweden
Occupational Overuse Syndrome	Australia
Repetitive Strain Injury	Australia, Canada, Netherlands
Work-Related Neck and Upper Limb Disorders; Work-Related Upper Limb Disorders	United Kingdom
Work-Related Musculoskeletal Disorders	World
Repetitive stress injury; Repetitive motion injuries	-

Table 1. WMSD designation (adapted from Nunes, 2003)

2.1 WMSD risk factors

The strong correlation between the incidence of WMSD and the working conditions is well known, particularly the physical risk factors associated with jobs e.g., awkward postures, high repetition, excessive force, static work, cold or vibration. Work intensification and stress and other psychosocial factors also seem to be factors that increasingly contribute to the onset of those disorders (EU-OSHA 2008; EU-OSHA 2011; HSE 2002; EUROFUND, 2007).

As referred WHO attributes a multifactorial etiology to WMSD, which means that these disorders appear as consequence of the worker exposure to a number of work related risk factors (WHO, 1985).

Besides risk factors related to work other risk factors contribute to its development, namely factors intrinsic to the worker and factors unrelated to work. A risk factor is any source or situation with the potential to cause injury or lead to the development of a disease. The variety and complexity of the factors that contribute to the appearance of these disorders explains the difficulties often encountered, to determine the best suited ergonomic intervention to be accomplished in a given workplace, to control them.

Moreover, despite all the available knowledge some uncertainty remains about the level of exposure to risk factors that triggers WMSD. In addition there is significant variability of individual response to the risk factors exposure.

The literature review and epidemiological studies have shown that in the genesis of the WMSD three sets of risk factors can be considered (Bernard, 1997; Buckle & Devereux, 1999; Nunes, 2009a):

- Physical factors - e.g., sustained or awkward postures, repetition of the same movements, forceful exertions, hand-arm vibration, all-body vibration, mechanical compression, and cold;
- Psychosocial factors - e.g., work pace, autonomy, monotony, work/rest cycle, task demands, social support from colleagues and management and job uncertainty;
- Individual factors - e.g., age, gender, professional activities, sport activities, domestic activities, recreational activities, alcohol/tobacco consumption and, previous WMSD.

In order to evaluate the possibility of an employee develop WMSD it is important to include all the relevant activities performed both at work and outside work. Most of the WMSD risk factors can occur both at work and in leisure time activities.

Risk factors act simultaneously in a synergistic effect on a joint or body region. Therefore to manage risk factors it is advisable and important to take into account this interaction rather than focus on a single risk factor. Due to the high individual variability it is impossible to estimate the probability of developing WMSD at individual level. As physicians usually say 'There are no diseases, but patients.'

2.1.1 Physical factors

A comprehensive review of epidemiological studies was performed to assess the risk factors associated with WMSDs (NIOSH, 1997). The review categorized WMSDs by the body part impacted including (1) neck and neck-shoulder, (2) shoulder, (3) elbow, (4) hand-wrist, and (5) back. The widely accepted physical or task-related risk factors include repetition, force, posture, vibration, temperature extremes, and static posture (NIOSH, 1997; McCauley Bush, 2011)

The physical risk factors are a subset of work related risk factors including the environment and biomechanical risk factors, such as posture, force, repetition, direct external pressure (stress per contact), vibration and cold. Another risk factor that affects all risk factors is duration. Since WMSD develop associated with joints, it is necessary that each of these risk factors is controlled for each joints of the human body. In Table 2 a compilation of physical risk factors by body area are presented.

2.1.2 Psychosocial factors

Psychosocial risk factors are non biomechanical risk factors related with work. The work-related psychosocial factors are subjective perceptions that workers have of the organizational factors, which are the objective aspects of how the work is organized, is supervised and is carried out (Hagberg et al., 1995). Although organizational and psychosocial factors may be identical, psychosocial factors include the worker emotional perception. Psychosocial risk factors are related with work content (eg, the work load, the task monotony, work control and work clarity), it organizational characteristics (for example, vertical or horizontal organizational structure), interpersonal relationships at work (e.g., relations supervisor-worker) and financial / economic aspects (eg, salary, benefits and equity) and social (e.g., prestige and status in society) (NIOSH, 1997). Psychosocial factors cannot be seen as risk factors that, by themselves, led to the development of WMSDs (Gezondheidsraad, 2000). However, in combination with physical risk factors, they can increase the risk of injuries, which has been confirmed by experience. Thus, if the psychological perceptions of the work are negative, there may be negative reactions of physiological and psychological stress. These reactions can lead to physical problems, such as muscle tension. On the other hand, workers may have an inappropriate behaviour at work, such as the use of incorrect working methods, the use of excessive force to perform a task or the omission of the rest periods required to reduce fatigue. Any these conditions can trigger WMSDs (Hagberg et al. 1995).

2.1.3 Individual or personal risk factors

The field of ergonomics does not attempt to screen workers for elimination as potential employees. The recognition of personal risk factors can be useful in providing training, administrative controls, and awareness. Personal or individual risk factors can impact the likelihood for occurrence of a WMSD (McCauley-Bell & Badiru, 1996a; McCauley-Bell & Badiru, 1996b). These factors vary depending on the study but may include age, gender, smoking, physical activity, strength, anthropometry and previous WMSD, and degenerative joint diseases (McCauley Bush, 2011).

Gender (McCauley Bush, 2011)

Women are three times more likely to have CTS than men (Women.gov, 2011). Women also deal with strong hormonal changes during pregnancy and menopause that make them more likely to suffer from WMSD, due to increased fluid retention and other physiological conditions. Other reasons for the increased presence of WMSDs in women may be attributed to differences in muscular strength, anthropometry, or hormonal issues. Generally, women are at higher risk of the CTS between the ages of 45 and 54. Then, the risk increases for both men and women as they age. Some studies have found a higher prevalence of some WMSDs in women (Bernard et al., 1997; Chiang et al., 1993; Hales et al., 1994), but the fact that more women are employed in hand-intensive jobs may account for the greater number of reported work-related MSDs among women. Likewise, (Byström et al., 1995) reported that men were more likely to have deQuervain's disease than women and attributed this to more frequent use of power hand tools. Whether the gender difference seen with WMSDs in some studies is due to physiological differences or differences in exposure is not fully understood.

	Neck and neck/shoulder	Shoulder	Elbow	Hand-wrist	Lower back	Lower limbs
Repetition	Repetitive neck movements Repeated arm or shoulder motions that generate loads to the neck-shoulder area (i.e., trapezius muscle)	Cyclical flexion, extension, abduction, or rotation of the shoulder joint	Cyclical flexion and extension of the elbow or cyclical Pronation, supination, extension, and flexion of the wrist that generates loads to the elbow–forearm region	Repetitive hand-finger or wrist movements (i.e., hand gripping) Wrist extension-flexion, ulnar-radial deviation, and supination or pronation Frequent repetitions have been defined as a cycle time <30 s or 50% of the task cycle spent performing the same activity (Silverstein et al., 1987).		Kneeling/ squatting Climbing stairs or ladders Heavy lifting Walking/standing
Posture	Extreme head or neck postures Static postures of the head and/or neck	When the arm is flexed, abducted, or extended, such that the angle between the torso and the upper arm increases	Repeated pronation, supination, flexion, or extension of the wrist, either singly or in combination with extension and flexion of the elbow	No neutral posture of the hand, wrist and/ or fingers-wrist flexion or extension, ulnar or radial deviation full hand grip, and pinch grip	Static - isometric positions where very little movement occurs, along with cramped or inactive postures Prolonged standing or sitting (sedentary work) No-neutral trunk postures (related to bending and twisting) in extreme positions or at extreme angles	

Table 2. WMSD physical risk factors by body area

	Neck and neck/shoulder	Shoulder	Elbow	Hand-wrist	Lower back	Lower limbs
Force	Forceful exertions involving the upper body that generates loads to the trapezius and neck muscles	Shoulder abduction, flexion, extension, or rotation to exert force	Strenuous activities involving the forearm extensors or flexors, which can generate loads to the elbow–forearm region	Forceful exertions performed by the hand, with or without a hand tool, during manipulative task activities	The physical stress that results from work done in transferring objects from one plane to another Forceful movements such as pulling, pushing, or other efforts	
Vibration		Low- or high-frequency vibration generally as a result of hand tools		Manual work involving vibrating power hand tools	Whole Body Vibration transferred to the body as a whole (in contrast to specific body regions), usually through a supporting system such as a seat or platform	
Cold				Workers may exert more force than necessary, affecting muscles, soft tissues, and joints May require gloves that have been shown to impact sensation thus leading to additional force exertion		

Table 2. (continued) WMSD physical risk factors by body area

To differentiate the effect of work risk factors from potential effects that might be attributable to biological differences, researchers must study jobs that men and women perform relatively equally.

Physical Activity (McCauley Bush, 2011)

Studies on physical fitness level as a risk factor for WMSDs have produced mixed results. Physical activity may cause injury. However, the lack of physical activity may increase susceptibility to injury, and after injury, the threshold for further injury is reduced. In construction workers, more frequent leisure time was related to healthy lower backs and severe low-back pain was related to less leisure time activity (Holmström et al., 1992). On the other hand, some standard treatment regimes have found that musculoskeletal symptoms are often relieved by physical activity. National Institute for Occupational Safety and Health (NIOSH, 1991) stated that people with high aerobic capacity may be fit for jobs that require high oxygen uptake, but will not necessarily be fit for jobs that require high static and dynamic strengths and vice versa.

Strength (McCauley Bush, 2011)

Epidemiologic evidence exists for the relationship between back injury and weak back strength in job tasks. Chaffin & Park (1973) found a substantial increase in back injury rates in subjects performing jobs requiring strength that was greater or equal to their isometric strength-test values. The risk was three times greater in weaker subjects. In a second longitudinal study, Chaffin et al. (1977) evaluated the risk of back injuries and strength and found the risk to be three times greater in weaker subjects. Other studies have not found the same relationship with physical strength. Two prospective studies of low-back pain reports (or claims) of large populations of blue collar workers (Battie et al., 1989; Leino, 1987) failed to demonstrate that stronger workers (defined by isometric lifting strength) are at lower risk for lowback pain claims or episodes.

Anthropometry (McCauley Bush, 2011)

Weight, height, body mass index (BMI) (a ratio of weight to height squared), and obesity have all been identified in studies as potential risk factors for certain WMSDs, particularly CTS and lumbar disc herniation. Vessey et al. (1990) found that the risk for CTS among obese women was double that of slender women. The relationship of CTS and BMI has been suggested to be related to increased fatty tissue within the carpal canal or to increased hydrostatic pressure throughout the carpal canal in obese persons compared with slender persons (Werner et al, 1994). Carpal tunnel canal size and wrist size has been suggested as a risk factor for CTS; however, some studies have linked both small and large canal areas to CTS (Bleecker, et al., 1985; Winn & Habes, 1990). Studies on anthropometric data are conflicting, but in general indicate that there is no strong correlation between stature, body weight, body build, and low back pain. Obesity seems to play a small but significant role in the occurrence of CTS.

Smoking (McCauley Bush, 2011)

Several studies have presented evidence that smoking is associated with low-back pain, sciatica, or intervertebral herniated disc (Finkelstein, 1995; Frymoyer et al.,1983; Kelsey et al., 1990; Owen & Damron, 1984; Svensson & Anderson, 1983); whereas in others, the relationship was negative (Frymoyer, 1991; Hildebrandt, 1987; Kelsey et al., 1990; Riihimäki et al., 1989). Boshuizen et al. (1993) found a relationship between smoking and back pain only in those occupations that required physical exertion. In this study, smoking was more clearly related to pain in the extremities than to pain in the neck or the back. Deyo & Bass (1989) noted that the prevalence of back pain increased with the number of pack-years of cigarette smoking and with the heaviest smoking level. Several explanations for the relationship have been proposed. One hypothesis is that back pain is caused by coughing from smoking.

Coughing increases the abdominal pressure and intradiscal pressure, thereby producing strain on the spine. Several studies have observed this relationship (Deyo & Bass, 1989; Frymoyer et al., 1980; Troup et al., 1987). Other theories include nicotine-induced diminished blood flow to vulnerable tissues (Frymoyer et al., 1983), and smoking-induced diminished mineral content of bone causing microfractures (Svensson & Andersson, 1983).

2.1.4 Interaction among risk factors

All risk factors interact among each other. For example, the stress felt by a worker may be influenced by the physical demands of the task, the psychological reaction to this requirement, or by both.

Once the requirement of the task reaches a high value, the worker may have stress reactions and biological and behavioral unsuitable reactions. As these reactions are more frequent and occur over an extended period they cause health problems. These health problems reduce the 'resistance' of individuals to cope with the subsequent demands of work, thus increasing the possibility of occurrence of WMSDs. As mentioned, the duration of exposure to risk factors is one of the parameters that must be taken into account when a risk assessment is performed. For example, the heuristic model dose-response (Figure 1) to cumulative risk factors in repetitive manual work, proposed by Tanaka McGlothlin, underlines the role of the duration of the activity in the development of musculoskeletal disorders of the hand / wrist (Tanaka & McGlothlin, 2001).

In the figure it's possible to observe the interaction of the following risk factors: force, repetition and wrist posture with exposure duration. In order to keep workers operating in a safe area an increase in exposure duration should be accompanying with a reduction of the other risk factors.

2.2 Models of WMSD pathophysiologic mechanisms

As mentioned before the term WMSD usually refers to disorders caused by a combination of risk factors that act synergistically on a joint or body region, over time. Until now the biological pathogenesis associated with the development of the majority of the WMSD is unknown. However several models have been proposed to describe the mechanisms that lead to the development of WMSDs, ie how different risk factors act on human body. See for instance the models proposed by (Armstrong et al. 1993; NRC, 1999; NRC & IOM 2001). Such models provide a guide to ergonomic interventions aiming to control the development of WMSDs.

The integrated model presented in Figure 2 combines the theories and models that accounted for the various possible mechanisms and pathways (Karsh, 2006). At the top of the model are the factors relating to workplace that determine exposure to WMSD risk factors ie, the work organization, the company socio-cultural context and the environment surrounding the workplace.

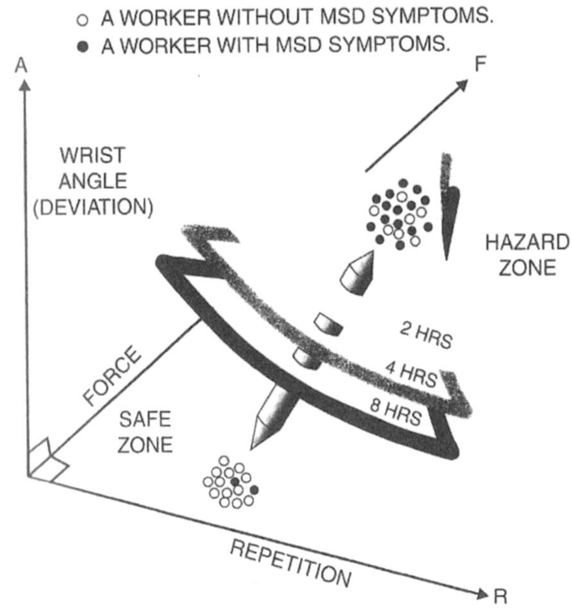

Fig. 1. Risk factors interaction (Tanaka & McGlothlin, 2001).

The mechanisms or pathways that can lead to development of WMSDs are numbered form 1 to 36 in the figure, and are explained below:

- '1' indicates that the social and cultural context of the organization influences the way work is organized;
- '2' shows that the social and cultural context of the organization may have a direct impact on psychological demands of work, through for example, the safety climate of the company;
- '3' and '4' represent the direct impact of work organization on the physical and psychological work demands, also indicating that the impact of the social / cultural context have in physical and psychological demands is mediated by the organization of work. Since the organization of work can be defined as the objective nature of the work, it determines the physical and psychological characteristics of work;
- '5' and '6' shows that the work environment, for example, lighting conditions, the noise, vibration or temperature may also influence directly the physical demands and psychological work demands. For example, reflections due to inadequate lighting conditions in a computer screen, can influence the posture adopted by the worker, in order not to be affected by the reflections;

- '7' is a reciprocal pathway between the physical and psychological demands of work, which indicates that these two types of requirements influence each other. For example, a job highly repetitive can influence the perception of low control over their activities that workers must have;
- '8' represents the direct impact of the physical work demands on physical strain. The mechanism by which this occurs and, consequently led to the development of WMSDs can be through over-exertion, accumulated charge, fatigue or changes in work style;
- '9' indicates the psychological tension generated by the physical demands;
- '10' shows that the psychological work demands can influence the psychological strain. These requirements may have a direct impact on psychological strain if the requirements cause psychological stress or anxiety. These influences may be due to changes in work style, increased muscle tension or psychological stress.
- '11' and '12' show that the physical and psychological demands of work can have a direct impact on the individual characteristics of workers, through mechanisms of adaptation such as improving their physical or psychological capacity;
- '13' is a reciprocal pathway that shows that the physical and psychological strains can influence each other. The psychological strain may impact physical strain by increasing the muscle tension, while the physical strain can influence psychological strain. Individual characteristics such as physical and psychological tolerance to fatigue and resistance to stress may moderate many of the above relationships. Thus:
- '14' physical capacity may moderate the relationship between the physical work demands and physical strain;
- '15' coping mechanisms may moderate the relationship between psychological work demands and physiological strain;
- '16' capacity and internal tolerances can impact the extent to which physical and psychological strain affect each other;
- '17' and '18' indicate that the physical and psychological strain can cause changes in physiological responses, which can provide new doses for other physical and psychological responses;
- '19', '20', '21', '35' indicate that the individual characteristics, the work organization, and the physical and psychological strain and the related physiological responses may have an impact in the detection of symptoms through mechanisms related to increased sensitivity;
- '22' represents the perception, identification and attribution of symptoms to 'something' by workers;
- '23' represents the fact that the symptoms can lead to WMSD diagnosis;
- '24' indicates that, even without symptoms, a WMSD may be present;
- '25', '26', '27' and '28' represent the fact that the existence of WMSDs may have effects on psychological and physical strain and / or the physical and psychological work demands, since the existence of a WMSD, can lead to modification in the way a worker performs his work, or increase psychological stress;
- '29', '30', '31' and '32' indicate that the mere presence of symptoms can lead a worker to modify the way he performs his work thus contributing to stress;
- '33' and '34' respectively indicate that the perception of symptoms or the presence of WMSDs can lead to redesign of the work, which has an impact on work organization.

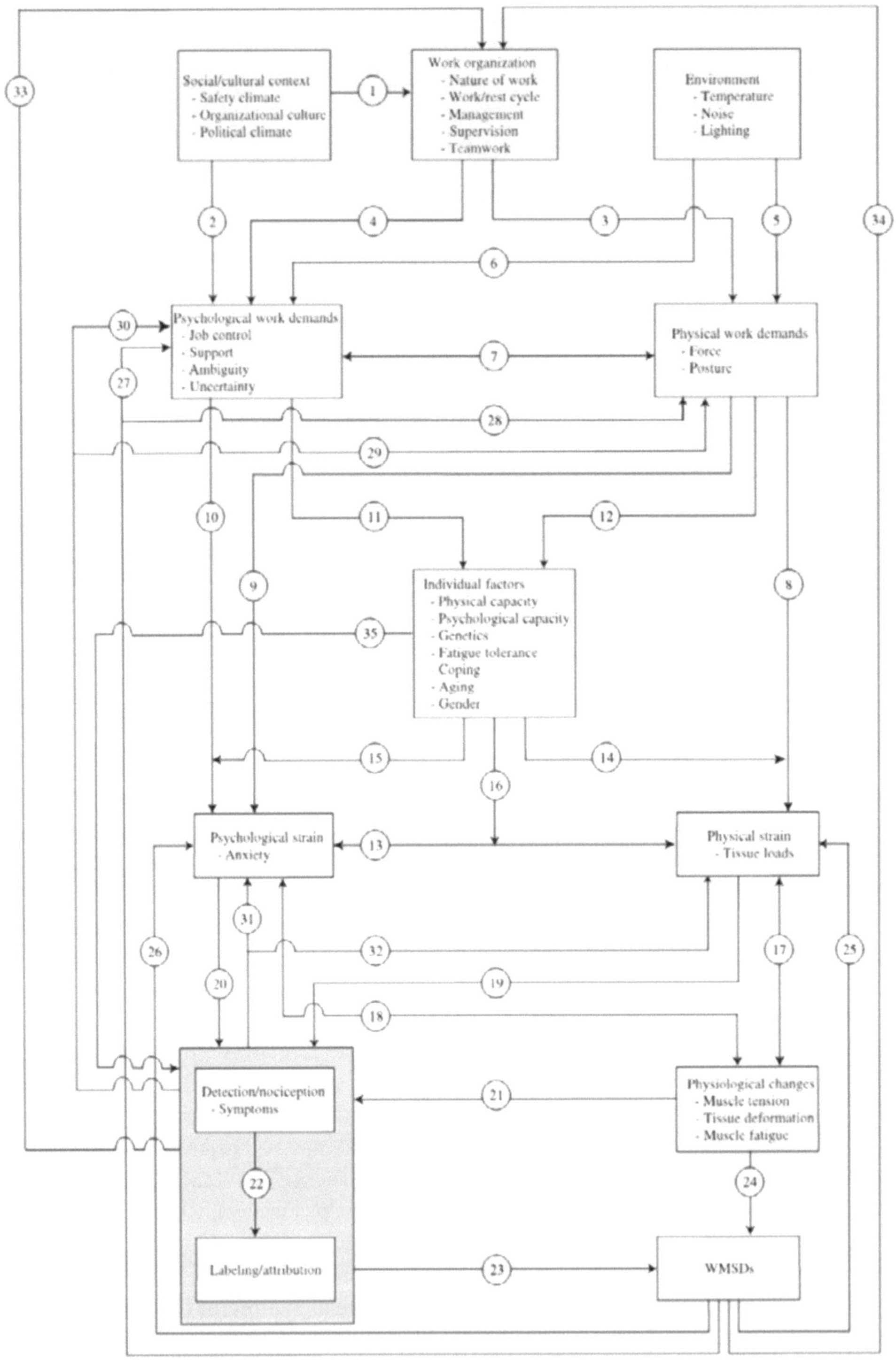

Fig. 2. WMSD integrated model (Karsh, 2006).

As referred non-professional activities can also contribute to the development of WMSD, thus we can add to this model a pathway '36' that represent sport or domestic activities. The pathway should impact the 'physical strain' box.

2.3 The most relevant WMSD and risk factors

WRMD are classified according to the affected anatomical structure (Putz-Anderson, 1988; Pujol, 1993; Hagberg et al., 1995):

- Tendon - include inflammation of the tendons and / or their synovial sheaths. These disorders are usually identify as tendonitis, which is the inflammation of tendons; tenosynovitis, which are injuries involving tendons and their sheaths, and synovial cysts, which are the result of lesions in the tendon sheath;
- Bursa – its inflammation is designated as bursitis;
- Muscles - muscles fatigue, such as, in Tension Neck Syndrome;
- Nerve - involve the compression of a nerve, such as the Carpal Tunnel Syndrome;
- Vascular - affects the blood vessels, as in vibration syndrome.

Table 3 shows the WMSDs that will be addressed in this document, organized according to region of the body where they occur and the anatomical structure affected.

The characterization of several WRMD is provided in the following paragraphs.

Tension Neck Syndrome

The Tension Neck Syndrome is a term that designates a set of muscle pain, accompanied by increased sensitivity and stiffness in the neck and shoulders, often registering muscle spasms. This syndrome is most common in women than in men. It has not been possible to determine whether this difference in incidence is due to genetic factors or exposure to different risk factors, both professional and unprofessional, characteristic of females, (Hagberg, et al., 1995). Epidemiological studies carried out by Bernard (NIOSH, 1997) revealed the existence of a causal relationship between the performance of highly repetitive work and the existence of this type of injury. The introduction of data in computer terminals is an example of a work situation where constrained arms and head postures occur during work.

Back Injuries (McCauley Bush, 2011)

The back is the most frequently injured part of the body (22% of 1.7 million injuries) (NSC, Accident Facts, 1990) with overexertion being the most common cause of these injuries. However, many back injuries develop over a long period of time by a repetitive loading of the discs caused by improper lifting methods or other exertions.

In fact, 27% of all industrial back injuries are associated with some form of lifting or manual material handling. These injuries are generally repetitive and result after months or years of task performance. Often injuries that appear to be acute are actually the result of long-term impact. The discs of the back vary in size, are round, rubber-like pads filled with thick fluid, which serve as shock absorbers. All the forces that come down the spine compress these discs, as a result of continuous and repetitive squeezing. In some instance disks can rupture and bulge producing pressure on the spinal nerve resulting in back pain.

Affected structure / Body part	WMSD Neck	Shoulder	Elbow	Wrist/ Hand	Lumbar area	Hip/ Thigh	Knee	Leg/ Foot
Tendons and sheaths		Shoulder Tendonitis	Epicondylitis	De Quervain Disease Tenosynovitis Wrist / Hand Synovial Cyst Trigger Finger		Piriformis Syndrome	Pre-patellar Tendonitis Shin splints Infra-patellar Tendonitis	Achilles Tendonitis
Bursa/ capsule		Shoulder Bursitis Frozen Shoulder (adhesive capsulitis)	Olecranon Bursitis					
Muscles	Tension Neck Syndrome					Trochanteritis		
Nerves	Cervical Spine Syndrome	Thoracic Outlet Syndrome	Radial Tunnel Syndrome Cubital Tunnel Syndrome	Carpal Tunnel Synd. Guyon's Canal Synd.	Low Back Pain	Piriformis Syndrome		
Blood vessels				Hand-Arm Syndrome (Raynaud Syndrome) Hypothenar Hammer Syndrome				Varicose veins Venous disorders
Bone/ cartillage						Sacroiliac Joint Pain	Pre-patellar Tendonitis	

Table 3. Most relevant WMSD by body part and affected anatomical structure (adapted from Nunes, 2003)

Carpal Tunnel Syndrome (McCauley Bush, 2011)

Perhaps the most widely recognized WMSD of the hand and forearm region is carpal tunnel syndrome (CTS), a condition whereby the median nerve is compressed when passing through the bony carpal tunnel (wrist). The carpal tunnel comprises eight carpal bones at the wrist, arranged in two transverse rows of four bones each. The tendons of the forearm muscles pass through this canal to enter the hand and are held down on the anterior side by fascia, called flexor and extensor retinacula, which are tight bands of tissue that protect and restrain the tendons as they pass from the forearm into the hand. If these transverse bands of fascia were not present, the tendons would protrude when the hand is flexed or extended (Spence, 1990). The early stages of CTS result when there is a decrease in the effective cross section of the tunnel caused by the synovium swelling and the narrowing of the confined space of the carpal tunnel. Subsequently, the median nerve, which accompanies the tendons through the carpal tunnel, is compressed and the resulting condition is CTS.

Early symptoms of CTS include intermittent numbness or tingling and burning sensations in the fingers. More advanced problems involve pain, wasting of the muscles at the base of the thumb, dry or shiny palms, and clumsiness. Many symptoms first occur at night and may be confined to a specific part of the hand. If left untreated, the pain may radiate to the elbows and shoulders.

Tendonitis (McCauley Bush, 2011)

Tendonitis, an inflammation of tendon sheaths around a joint, is generally characterized by local tenderness at the point of inflammation and severe pain upon movement of the affected joint. Tendonitis can result from trauma or excessive use of a joint and can afflict the wrist, elbow (where it is often referred to as 'tennis elbow'), and shoulder joints.

Tenosynovitis (McCauley Bush, 2011)

Tenosynovitis is a repetition-induced tendon injury that involves the synovial sheath. The most widely recognized tenosynovitis is deQuervain's disease. This disorder affects the tendons and sheaths on the side of the wrist and at the base of the thumb.

Intersection Syndrome and deQuervain's Syndrome (McCauley Bush, 2011)

Intersection syndrome and deQuervain's syndrome occur in hand-intensive workplaces.

These injuries are characterized by chronic inflammation of the tendons and muscles on the sides of the wrist and the base of the thumb. Symptoms of these conditions include pain, tingling, swelling, numbness, and discomfort when moving the thumb.

Trigger Finger (McCauley Bush, 2011)

If the tendon sheath of a finger is aggravated, swelling may occur. Sufficient amounts of swelling may result in the tendon becoming locked in the sheath. At this point, if the person attempts to move the finger, the result is a snapping and jerking movement.

This condition is called trigger finger. Trigger finger occurs to the individual or multiple fingers and results when the swelling produces a thickening on the tendon that catches as it runs in and out of the sheath. Usually, snapping and clicking in the finger arises with this disorder. These clicks manifest when one bends or straightens the fingers (or thumb). Occasionally, a digit will lock, either fully bent or straightened.

Ischemia (McCauley Bush, 2011)

Ischemia is a condition that occurs when blood supply to a tissue is lacking. Symptoms of this disorder include numbness, tingling, and fatigue depending on the degree of ischemia, or blockage of peripheral blood vessels. A common cause of ischemia is compressive force in the palm of the hand.

Vibration Syndrome (McCauley Bush, 2011)

Vibration syndrome is often referred to as white finger, dead finger, or Raynaud's phenomenon. These conditions are sometimes referred to as hand arm vibration (HAV) syndrome. Excessive exposure to vibrating forces and cold temperatures may lead to the development of these disorders. It is characterized by recurrent episodes of finger blanching due to complete closure of the digital arteries.

Thermoregulation of fingers during prolonged exposure to cold is recommended, as low temperatures reduce blood flow to the extremities and can exacerbate this condition.

Thoracic Outlet Syndrome (McCauley Bush, 2011)

Thoracic outlet syndrome (TOS) is a term describing the compression of nerves (brachial plexus) and/or vessels (subclavian artery and vein) to the upper limb.

This compression occurs in the region (thoracic outlet) between the neck and the shoulder. The thoracic outlet is bounded by several structures: the anterior and middle scalene muscles, the first rib, the clavicle, and, at a lower point, by the tendon of the pectoralis minor muscle. The existence of this syndrome as a true clinical entity has been questioned, because some practitioners suggest that TOS has been used in error when the treating clinician is short on a diagnosis and unable to explain the patient's complaints. Symptoms of TOS include aching pain in the shoulder or arm, heaviness or easy fatigability of the arm, numbness and tingling of the outside of the arm or especially the fourth and fifth fingers, and finally swelling of the hand or arm accompanied by finger stiffness and coolness or pallor of the hand.

Ganglion Cysts (McCauley Bush, 2011)

Ganglion is a Greek word meaning 'a knot of tissue.' Ganglion cysts are balloon like sacs, which are filled with a jelly-like material. The maladies are often seen in and around tendons or on the palm of the hand and at the base of the finger. These cysts are not generally painful and with reduction in repetition often leave without treatment.

Lower limbs WMSD

Lower limb WMSD are currently a problem in many jobs, they tend to be related with disorders in other areas of the body. The epidemiology of these WMSD has received until now modest awareness, despite this there is appreciable evidence that some activities (e.g., kneeling/ squatting, climbing stairs or ladders, heavy lifting, walking/standing) are causal risk factors for their development. Other causes for acute lower limb WMSD are related with slip and trip hazards (HSE, 2009). Despite the short awareness given to this type of WMSD they deserve significant concern, since they often are sources of high degrees of immobility and thereby can substantially degrade the quality of life (HSE, 2009). The most common lower limb WMSD are (HSE, 2009):

- Hip/thigh conditions - Osteoarthritis (most frequent), Piriformis Syndrome, Trochanteritis, Hamstring strains, Sacroiliac Joint Pain;
- Knee / lower leg - Osteoarthritis, Bursitis, Beat Knee/Hyperkeratosis, Meniscal Lesions, Patellofemoral Pain Syndrome, Pre-patellar Tendonitis, Shin Splints, Infra-patellar Tendonitis, Stress Fractures;
- Ankle/foot - Achilles Tendonitis, Blisters, Foot Corns, Halux Valgus (Bunions), Hammer Toes, Pes Traverse Planus, Plantar Fasciitis, Sprained Ankle, Stress fractures, Varicose veins, Venous disorders.

Non-specific WMSD

Non-specific WMSD are musculoskeletal disorders that have ill-defined symptoms, i.e. the symptoms tend to be diffuse and non-anatomical, spread over many areas: nerves, tendons and other anatomical structures (Ring et al. 2005). The symptoms involve pain (which becomes worse with activity), discomfort, numbness and tingling without evidence of any discrete pathological condition.

2.4 Summary of WMSD, symptoms and occupational risk factors

The assessment of WMSD's can be done using multiple checklists, subjective and objective assessments. An efficient approach is to identify occupational risk factors and make efforts to remove them from task. Where the risk factors cannot be removed the impact should be reduced and mitigation strategies employed to reduce the likelihood for injury. Administrative controls such as more frequent rest breaks, task sharing or rotation between jobs. Table 4 provides a summary of common WMSDs, symptoms and risk factors.

Identified disorders, occupational risk factors and symptoms		
Disorders	**Occupational risk factors**	**Symptoms**
Tendonitis/tenosynovitis	Repetitive wrist motions Repetitive shoulder motions Sustained hyper extension of arms Prolonged load on shoulders	Pain, weakness, swelling, burning sensation or dull ache over affected area
Epicondylitis (elbow tendonitis)	Repeated or forceful rotation of the forearm and bending of the wrist at the same time	Same symptoms as tendonitis
Carpal tunnel syndrome	Repetitive wrist motions	Pain, numbness, tingling, burning sensations, wasting of muscles at base of thumb, dry palm
DeQuervain's disease	Repetitive hand twisting and forceful gripping	Pain at the base of thumb
Thoracic outlet syndrome	Prolonged shoulder flexion Extending arms above shoulder height Carrying loads on the shoulder	Pain, numbness, swelling of the hands
Tension neck syndrome	Prolonged restricted posture	Pain

Table 4. Work Related Musculoskeletal Disorders, Symptoms and Risk Factors (Canadian Centre for Occupational Health and Safety, 2011)

3. Procedure for workplace analysis and design (McCauley Bush, 2011)

Job analysis, risk factor assessment, and task design should be conducted to identify potential work-related risks and develop engineering controls, administrative controls, and personal protective resources to mitigate the likelihood of injuries. According to American National Standards Institute (ANSI), this can be accomplished with the following steps (Karwowski & Marras, 1998):

- Collect pertinent information for all jobs and associated work methods.
- Interview a representative sample of affected workers.
- Breakdown a job into tasks or elements.
- Description of the component actions of each task or element.
- Measurement and qualification or quantification of WMSDs (where possible).
- Identification of risk factors for each task or element.
- Identification of the problems contributing to the risk factors.
- Summary of the problem areas and needs for intervention for all jobs and associated new work methods.

These steps can be executed utilizing any combination of scientifically based assessment techniques including surveys, electronic measurement equipment, software tools, and analysis approaches.

4. Ergonomic tools for assessing WMSD risk factors

A diversity of ergonomic tools has been developed in order to help in the identification of WMSD risk factors and assessing the risk present on workstations. Some of the tools already developed are, for instance, OWAS (Karhu et al. 1977) (and the associated software WinOWAS (Tiilikainen, 1996)), RULA (McAtamney & Corlett, 1993), Strain Index (Moore and Garg 1995), NIOSH (Waters et al. 1993), (NIOSH, 1994), OCRA (Occhipinti, 1998), (Occhipinti & Colombini 2007), Quick Exposure Check (Li & Buckle, 1999), a fuzzy predictive model developed by McCauley Bell (McCauley-Bell & Badiru, 1996a) and FAST ERGO_X (Nunes, 2009a). The two systems developed by the chapter authors will be presented below.

4.1 Fuzzy risk predictive model

The development of quantitative model for industry application was the focus of research that produced the McCauley Bush and Badiru approach to prediction of WMSD risk. This model is intended for use as a method obtain the likelihood for WMSD risk for a specific individual performing a task at a given organization. The research identified three broad categories (modules) for WMSD risk factors as task-related, personal-related and organizational-related classifications. Within each of these categories, additional factors were identified. The items identified as risk factors for each of the three modules (task, personal and organizational) were evaluated for relative significance. The relative significance (priority weights) for the risk factors in the task-related and personal modules are listed in Tables 5 and Table 8, respectively. The levels of existence for each factor within the task related category is also shown in Table 6.

Ranking	Factor	Relative Weight
1	Awkward joint posture	0.299
2	Repetition	0.189
3	Hand tool use	0.180
4	Force	0.125
5	Task duration	0.124
6	Vibration	0.083

Table 5. AHP Results: Task-Related Risk Factors

	Posture	Repetition	Hand Tool	Force	Task Duration	Vibration
High	1.0	1.0	1.0	1.0	1.0	1.0
Medium	0.5	0.5	0.5	0.5	0.5	0.5
Low	0.2	0.2	0.2	0.2	0.2	0.2
None	0.0	0.0	0.0	0.0	0.0	0.0

Table 6. Levels of Existence for each factor

In the evaluation the organizational risk factors, equipment was the most significant factor. The term equipment refers to the degree of automation for the machinery being used in the task under evaluation. The relative significance and for each of the risk factors is listed in Table 6. This module evaluated the impact of seven risk factors. However, upon further analysis and discussion, the awareness and ergonomics program categories were combined because according to the experts and the literature, one of the goals of an ergonomics program is to provide awareness about the ergonomic risk factors present in a workplace.

Ranking	Factor	Relative Weight
1	Previous CTD	0.383
2	Hobbies and habits	0.223
3	Diabetes	0.170
4	Thyroid problems	0.097
5	Age	0.039
6	Arthritis or Degenerative Joint Disease (DJD)	0.088

Table 7. AHP Results: Personal Risk Factors

	Previous CTD	Hobbies & Habits	Diabetes	Thyroid Condition	Age	Arthritis or DJD
High	1.0	1.0	1.0	1.0	1.0	1.0
Medium	0.5	0.5	0.5	0.5	0.5	0.5
Low	0.2	0.2	0.2	0.2	0.2	0.2
None	0.0	0.0	0.0	0.0	0.0	0.0

Table 8. Levels of Existence for Personal Risk Factors

Ranking	Factor	Relative Weight
1	Equipment	0.346
2	Production rate/layout	0.249
3	Ergonomics program	0.183
4	Peer influence	0.065
5	Training	0.059
6	CTD level	0.053
7	Awareness	0.045

Table 9. AHP Results: Organizational Risk Factors

	Equipment	Production rate/layout	Ergonomics program	Peer influence	Training	CTD level	Awareness
High	1.0	1.0	1.0	1.0	1.0	1.0	1.0
Medium	0.5	0.5	0.5	0.5	0.5	0.5	0.5
Low	0.2	0.2	0.2	0.2	0.2	0.2	0.2
None	0.0	0.0	0.0	0.0	0.0	0.0	0.0

Table 10. Levels of Existence for Organizational Risk Factors

After the factors within the categories (or modules) were compared, analytic Hierarchy Processing (AHP) analysis was conducted to determine the relative significance of each of the modules: task, personal and organizational characteristics. The relative significance (priority weights) obtained for the task, personal, and organizational characteristics are listed in Table 11. The task characteristics module received a relative weight of 0.637. The personal characteristics module had a relative weight of 0.258, less than half of the relative weight of the task characteristics module. Finally, the organizational characteristics module received the smallest relative weight, 0.105.

Ranking	Module	Relative Weight
1	Task	0.637
2	Personal	0.258
3	Organizational	0.105

Table 11. AHP Results: Module Risk Comparison

Determination of Aggregate Risk Level

After the linguistic risk and the relative significance are generated an aggregated numeric value is obtainable. Equation 1 represents the model for the calculation of the numeric risk value for the task module. In Equation 1, the w_i values represent the numeric values obtained from the user inputs for each of the six risk factors and the a_j values represent the relative significance or factor weight obtained from the AHP analysis. The numeric risk levels for the personal and organizational characteristics are represented by Equations 2 and 3, respectively. Likewise, the values of x_i and y_i represent the user inputs while, the b_i and c_j

values represent the AHP weights for the task and organizational characteristics, respectively. These linear equations are based on Fuzzy Quantification Theory I (Terano et al, 1987). The objective of Theory I is to find the relationships between the qualitative descriptive variables and the numerical object variables in the fuzzy groups. An alternative to this approach is to use CTD epidemiological data to establish the regression weights rather than the relative weights were derived from the AHP analysis with the experts. However, the lack of availability of comprehensive data for a regression model prevented the application of regression analysis. The resulting equations represent the numeric risk levels for each category.

Task-Related Risk:

$$R_1 = F(T) = a_1 w_1 + a_2 w_2 + a_3 w_3 + a_4 w_4 + a_5 w_5 + a_6 w_6 \tag{1}$$

Personal Risk:

$$R_2 = F(P) = b_1 x_1 + b_2 x_2 + b_3 x_3 + b_4 x_4 + b_5 x_5 + b_6 x_6 \tag{2}$$

Organizational Risk:

$$R_3 = F(O) = c_1 y_1 + c_2 y_2 + c_3 y_3 + c_4 y_4 + c_5 y_5 + c_6 y_6 \tag{3}$$

Interpretation of Results

The numeric risk values obtained from each of the modules and the weights obtained from the AHP analysis were used to calculate the overall risk level. This value indicates the risk of injury for the given person, on the evaluated task for the workplace under evaluation (Equation 4). The following equation was used to quantify the comprehensive risk of injury is a result of all three categories:

Comprehensive Risk:

$$Z = d_1 R_1 + d_2 R_2 + d_3 R_3 \tag{4}$$

where,

Z = overall risk for the given situation,
R_1 = the risk associated with the task characteristics,
d_1 = weighting factor for the task characteristics,
R_2 = the risk associated with the personal characteristics,
d_2 = weighting factor for the personal characteristics,
R_3 = the risk associated with the organizational characteristics,
d_3 = weighting factor for the organizational characteristics.

The weighting factors (d_1, d_2, d_3) represent the relative significance of the given risk factor category's contribution to the likelihood of injury. These factors were determined through the AHP analysis. The numeric risk levels obtained from the previous equations exist on the interval [0,1]. On this interval 0 represents 'no risk of injury' and 1 represents 'extreme risk of injury'. The interpretation and categorization is shown in Table 12.

Numeric Risk Level	Expected Amount of Risk Associated with Numeric Value
0.00 - 0.20	Minimal risk: Individual should not be experiencing any conditions that indicated musculoskeletal irritation
0.21 - 0.40	Some risk: may be in the very early stages of CTD development. Individual may experience irregular irritation but is not expected to experience regular musculoskeletal irritation
0.41 - 0.60	Average risk: Individual may experience minor musculoskeletal irritation on a regular but not excessive irritation
0.61 - 0.80	High risk: Individual is expected to be experiencing regular minor or major musculoskeletal irritation
0.81 - 1.00	Very high risk: Individual is expected to presently experience ongoing or regular musculoskeletal irritation and/or medical correction for the condition

Table 12. Interpretation and Categorization of aggregate risk levels

4.2 FAST ERGO_X

FAST ERGO_X is a system whose aim is to assist Occupational Health and Safety professionals in the identification, assessment and control of ergonomic risks related with the development of WMSD. It was designed to identify, evaluate and control the risk factors due to ergonomic inadequacies existing in the work system (Nunes, 2009a). This method was devedelop in Faculdade de Ciências e Tecnologia da Universidade Nova de Lisboa, Portugal.

As referred before despite all the available knowledge there remains some uncertainty about the precise level of exposure to risk factors that triggers WMSD. In addition there is significant variability of individual response to the risk factors exposure. Aware that there was yet room for use of alternative approaches and the development of new features, and recognizing the adequacy of applying fuzzy expert systems for dealing with the uncertainty and imprecision inherent to the factors considered in an ergonomic analysis, the fuzzy expert system model for workstation ergonomic analysis, named ERGO_X and a first prototype were developed (Nunes et al. 1998), (Nunes, 2006). The ERGO_X method of workstation ergonomic analysis was subject to a Portuguese patent (Nunes, 2009b). FAST ERGO_X application was then developed based on the ERGO_X model, therefore FAST ERGO_X is a fuzzy expert system. This is an innovative approach that uses Artificial Intelligence concepts. This approach presents some advantages over the classical methods commonly used.

Based on objective and subjective data, the system evaluates the risk factors present in workplaces that can lead to the development of WMSD, and presents the findings of the evaluation. The system also presents recommendations that users can follow to eliminate or at least reduce the risk factors present in the work situation.

The FAST ERGO_X has the following features:

- data collection - supports the user to collect data, directing the collection and the filling of the data, according to the settings of analysis defined by the user and characteristics of the workstations and tasks under analysis;
- risk factors assessment - performs the assessment of risk factors present on the workplace, synthetizing the elements of analysis, presenting the conclusions in graphical or text formats;
- explanations presentation - provides explanations about the results obtained in the ergonomics analysis allowing an easy identification of individual risk factors that contributed to the result displayed;
- advisement - advises corrective or preventive measures to apply to the work situations, since the knowledge base includes a set of recommendations in HTML format, with hyperlinks that enable the navigation to a set of relevant topics related to the issues addressed (for example, risk factors, potential consequences, preventive measures or good practice references).

The use of FAST ERGO_X comprehends three main phases: analysis configuration, data collection and data analysis. These phases are depicted in Figure 3.

The use of the FAST ERGO_X is very flexible. On one hand, because it allows the use of objective and subjective data, separately or combined; on the other hand because it can be used on portable computers, which makes its utilization possible in situ either to collect data, to present the results and to support any decision-making that may be required, for instance due to the need of corrective interventions.

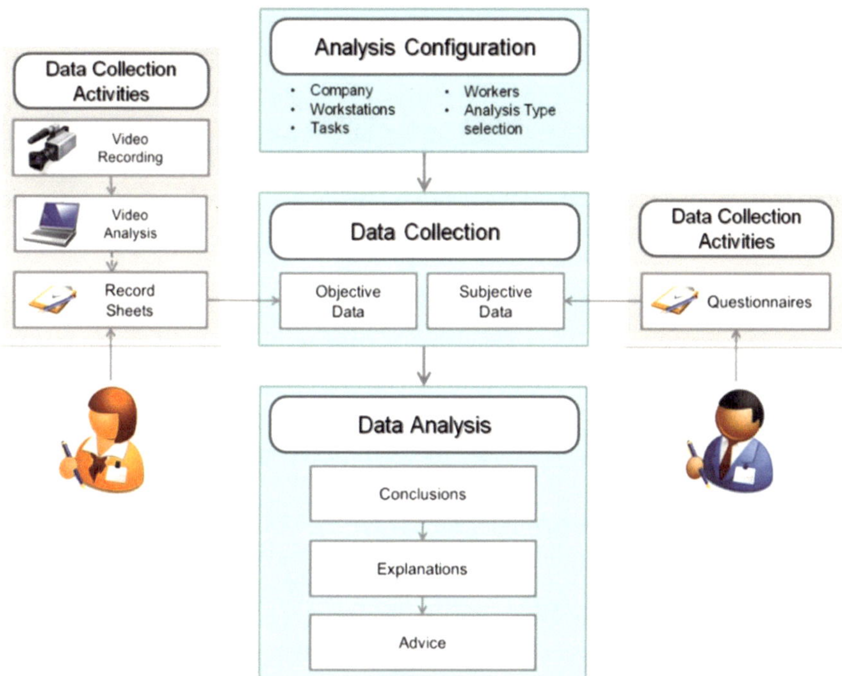

Fig. 3. Activities performed on the analysis of a work situation (Nunes, 2009a).

The forecast capability of the evaluation model allows the use of the system as a WMSD prevention tool creating the opportunity to act on identified risk factors, avoiding the WMSD associated costs and pains.

Finally, FAST ERGO_X can also be used as a tool to promote participatory ergonomics. For instance, the software and the media used for the analysis of the work situations (e.g., video recordings) can be used to support the training of workers in the field of Occupational Safety and Health. This can be achieved either by using the knowledge repository compiled on the knowledge base, by discussing the results of analyses carried out, or by proceeding to critical reviews of the videos collected for the analysis of work situations. Workers' awareness is a key success factor for the reduction of potentially risky behaviours, the identification of inadequate situations, and the development of solutions that help the prevention of WMSD. An example of application can be found in (Nunes, 2009a).

4.3 Additional screening methods for WMSD

Several methods have been developed to screen for, diagnose and treat musculoskeletal disorders. A few examples of screening approaches are discussed below.

4.3.1 Tinel's sign

Jules Tinel, a French neurologist, developed Tinel's Sign in 1915. He noted that after an injury, tapping of the median nerve resulted in a tingling sensation (paresthesia) in the first three and a half digits. Tinel's Sign was not originally associated with carpal tunnel syndrome; it was not until 1957 that George Phalen recognized that Tinel's Sign could be used to diagnose carpal tunnel syndrome (Urbano, 2000). Tinel's method is among the simplest and oldest screening approaches however, the application of this approach requires knowledge in ergonomics and an understanding of the technique. This subjective assessment technique requires input from the subject and can be a useful initial assessment tool however it should be coupled with additional ergonomic assessment tools.

4.3.2 Phalen's test

George S. Phalen, an American hand surgeon, studied patients with carpel tunnel syndrome and recognized that Tinel's Sign could be used to diagnose carpel tunnel syndrome, described it as 'a tingling sensation radiating out into the hand, which is obtained by light percussion over the median nerve at the wrist' (Urbano, 2000). Additionally, Phalen developed a wrist flexing test to diagnose carpel tunnel syndrome. To perform the Phalen's test, the patient should place their elbows on a table, placing the dorsal surfaces of the hands against each other for approximately 3 minutes. The patient should perform this maneuver with the wrists falling freely into their maximum flexion, without forcing the hands into flexion. Patients who have carpel tunnel syndrome will experience tingling or numbness after 1 to 2 minutes, whereas a healthy patient without carpel tunnel syndrome can perform the test for 10 or more minutes before experiencing tingling or numbness (Urbano, 2000).

4.3.3 Durkan test or carpal compression test

In 1991, John A. Durkan, an American orthopaedic surgeon, developed the carpal compression test. In a study of 31 patients with carpal tunnel syndrome, he found that this compression test was more sensitive than the Tinel's or Phalen's tests (Durkan, 1991). The carpal compression test involves directly compressing the median nerve using a rubber atomizer-bulb connected to a pressure manometer from a sphygmomanometer. This direct compression uses a pressure of 150 millimeters of mercury for 30 seconds. The occurrence of pain or paresthesia (tingling) indicates the presence of carpal tunnel syndrome. Durkan also identified an alternate method of performing the compression test by having the examiner apply even pressure with both thumbs to the median nerve in the carpal tunnel (Durkan, 1991).

4.3.4 Vibrometry testing

Vibrometry testing uses sensory perception to determine presence of carpal tunnel syndrome. To utilize this technique, the middle finger is placed on a vibrating stylus. While the evaluator manipulates vibration by altering the frequencies, the patient indicates whether or not they can detect the stylus vibrating. In theory, those with patients with carpal tunnel syndrome will be less sensitive to vibration. However, the effectiveness of vibrometry testing is debated with some studies as it has not conclusively been able to successfully identify carpal tunnel syndrome (Neese & Konz, 1993; Jetzer, 1991), while others show vibrometry testing to be inconclusive (Werne et al., 1994; White et al., 1994).

4.3.5 Nervepace electroneurometer device

Nervepace Electroneurometer is an objective method to test motor nerve conduction and infer the presence of carpal tunnel syndrome. Electrodes for surface stimulation are placed on the median nerve, approximately 3 cm proximal to the distal wrist flexor crease, while recording electrodes are placed on the muscles of the hand. The evaluator then adjusts the stimulus applied to the median nerve until a motor response is detected. The device records the latency between the stimulus and the response times, which the evaluator can use to determine the presence of carpal tunnel syndrome. However, studies have shown that the device can be made ineffective due to skin thickness (callous), peripheral neuropathy, or severe carpal tunnel syndrome. In addition, the American Association of Electrodiagnostic Medicine deemed the Nervepace Electroneurometer as 'flawed,' 'experimental,' and 'not an effective substitute for standard electrodiagnostic studies in clinical evaluation of patients with suspected CTS' (David et al., 2003; Pransky et al., 1997).

5. Conclusion

The objective of this chapter is to provide an introduction to WMSDs, associated risk factors and tools that can be useful in reducing the risks of these injuries. Application of ergonomic, biomechanical and engineering principles can be effective in reducing the risks and occurrence of WMSD. Epidemiological data has demonstrated that occupational risk factors such as awkward postures, highly repetitive activities or handling heavy loads are among the risk factors that studies have shown to damage the bones, joints, muscles, tendons, ligaments, nerves and blood vessels, leading to fatigue, pain and WMSDs. The effective design of ergonomic tools, equipment, processes and work spaces can have a tremendous effect on the risks and occurrence of WMSD.

6. References

Armstrong, T. J., Buckle, P., Fine, L. J., Hagberg, M., Jonsson, B., Kilbom, A., Kuorinka, I. A. A., Silverstein, B. A., Sjogaard, G., Viikari-Juntura, E. (1993). A conceptual model for work-related neck and upper-limb musculoskeletal disorders. *Scandinavian Journal of Work, Environment and Health,* Vol. 19, No. 2, pp. 73-84

Battié, M.C., Bigos, S.J., Fisher, L.D., Hansson, T.H., Jones, M.E., Wortley, M.D. (1989). Isometric lifting strength as a predictor of industrial back pain. *Spine* Vol. *14*, No. 8, pp. 851–856

Bernard, B., Ed. (1997). *Musculoskeletal Disorders and Workplace Factors: A critical review of epidemiologic evidence for work-related musculoskeletal disorders of the neck, upper extremity, and low back,* National Institute for Occupational Safety and Health

Bleecker, M.L., Bohlman, M., Moreland, R., Tipton, A. (1985). Carpal tunnel syndrome: role of carpal canal size. *Neurology,* Vol. *35*, No. 11, pp.1599–1604

Boshuizen, H.C., Verbeek, J.H.A.M., Broersen, J.P.J., Weel, A.N.H. (1993). Do smokers get more back pain? *Spine,* Vol. *18*, No. 1, pp. 35–40

Buckle, P., Devereux, J. (1999). *Work-related Neck and Upper Limb Musculoskeletal Disorders,* European Agency for Safety and Health at Work-2000

Canadian Occupational Safety and Health (2011). *Work Related Musculoskeletal Disorders, Symptoms and Risk Factors.* Available at http://www.ccohs.ca/oshanswers/diseases/rmirsi.html (accessed April, 2011)

Chaffin, D.B., Park, K.S. (1973). A longitudinal study of low-back pain as associated with occupational weight lifting factors. *American Industrial Hygiene Association Journal,* Vol. *34*, pp. 513–525

Chaffin, D.B., Herrin, G.D., Keyserling, W.M., Foulke, J.A. (1977). *Preemployment strength testing in selecting workers for materials handling jobs.* Cincinnati, OH: U.S. Department of Health, Education, and Welfare, Health Services and Mental Health Administration, National Institute for Occupational Safety and Health, DHEW (NIOSH) Publication No. 77–163

David W., Chaudhry V, Dubin A., Shields R. Jr. (2003). Literature review: Nervepace digital electroneurometer in the diagnosis of carpal tunnel syndrome. *Muscle Nerve,* Vol. 27, No. 3, pp. 378-385

Chiang H-C, Yin-Ching K, Chen S-S, Hsin-Su Y, Trong-Neng W, Chang P-Y. (1993). Prevalence of shoulder and upper-limb disorders among workers in the fish-processing industry. *Scandinavian Journal of Work Environment and Health,* Vol. 19, pp. 126-131

Deyo, R.A., Bass, J.E. (1989). Lifestyle and low back pain: the influence of smoking and obesity. *Spine,* Vol. *14*, No. 5, pp. 501–506

Durkan, J. A. (1991). A New Diagnostic Test for Carpal Tunnel Syndrome. *The Journal of Bone and Joint Surgery,* Vol. 73, No. 4, pp. 535-538

EU-OSHA (2008). *Work-related musculoskeletal disorders: Prevention report.*Available at: http://osha.europa.eu/en/publications/reports/en_TE8107132ENC.pdf European Agency for Safety and Health at Work

EU-OSHA (2010). *OSH in figures: Work-related musculoskeletal disorders in the EU – Facts and figures*. Available at: http://osha.europa.eu/en/publications/reports/TERO09009ENC, European Agency for Safety and Health at Work

EU-OSHA (2011). *Musculoskeletal Disorders: General questions*. Available at: http://osha.europa.eu/en/faq/frequently-asked-questions, European Agency for Safety and Health at Work

EUROFOUND (2007). *Musculoskeletal disorders and organisational change. Conference report*. Lisbon, European Foundation for the Improvement of Living and Working Conditions

EUROFOUND (2010). *Living and working in Europe*, European Foundation for the Improvement of Living and Working Conditions

EUROFOUND (2011). *Fifth European Working Conditions survey - 2010*. Available at www.eurofound.europa.eu/surveys/ewcs/index.htm

Finkelstein, M.M. (1995). Back pain and parenthood. *Occupational Environmental Medicine*, Vol. *52*, No. 1, pp.51–53

Frymoyer, J.W., Pope, M.H., Costanza, M.C., Rosen, J.C., Goggin. J.E., Wilder, D.G. (1980). Epidemiologic studies of low-back pain. *Spine*, Vol. 5, No. 5, pp. 419–423

Frymoyer, J.W., Pope, M.H., Clements, J.H. (1983). Risk factors in low back pain. *The Journal of Bone & Joint Surgery*, Vol. 65, No 2, pp. 213-218

Frymoyer, J.W., Cats-Baril, W.L. (1991). An overview of the incidence and costs of low back pain. *Orthop Clin North Am*, Vol. 22, pp. 262–271

Gezondheidsraad. (2000). *RSI*, Health Council of the Netherlands

Hagberg M., Silverstein B., Wells R., Smith M., Hendrick H., Carayon, P. (1995). *Work-related musculoskeletal disorders (WMSDs): a reference book for prevention*. London, England: Taylor and Francis

Hales, T.R., Sauter, S.L., Peterson, M.R., Fine, L.J., Putz-Anderson, V., Schleifer, L.R., et al. (1994). Musculoskeletal disorders among visual display terminal users in a telecom-munications company. *Ergonomics*, Vol. *37*, No. 10, pp. 1603–1621

Hildebrandt, V.H. (1987). A review of epidemiological research on risk factors of low back pain. In: Buckle P. (Ed.) *Musculoskeletal disorders at work*. Proceedings of a conference held at the University of Surrey, Guilford, April, 1987. London, England: Taylor and Francis, pp. 9–16

Holmström, E.B., Lindell, J., Moritz, U. (1992). Low back and neck/shoulder pain in construction workers: Occupational workload and psychosocial risk factors. Part 2: Relationship to neck and shoulder pain. *Spine*, Vol. *17*, No. 6, pp. 672-677

HSE (2002). *Upper limb disorders in the workplace*. Available at http://www.hseni.gov.uk/upper_limb_disorders_in_the_workplace.pdf. Health and Safety Executive

HSE (2009). *Lower limb MSD. Scoping work to help inform advice and research planning*. Available at http://www.hse.gov.uk/research/rrpdf/rr706.pdf. Health and Safety Executive

Jetzer, T. C. (1991). Use of Vibration Testing in the Early Evaluation of Workers with Carpal Tunnel Syndrome. *Journal of Occupational Medicine*, Vol. 23, No 3, pp. 117-120

Karhu, O., Kansi, P., Kuorinka, I. (1977). Correcting working postures in industry: a practical method for analysis. *Applied Ergonomics*, Vol. 8, pp. 199-201

Karwowski, W., Marras, W.S. (Eds.) (1998). *The occupational ergonomics handbook.* Boca Raton, FL: CRC Press

Karsh, B. (2006). Theories of work-related musculoskeletal disorders: implications for ergonomic interventions. *Theoretical Issues in Ergonomic Science*, Vol. 7, No. 1, pp. 71-88

Kelsey, C.A., Mettler, F.A. (1990). Flexible protective gloves: the Emperor's new clothes? *Radiology*, Vol. 174, No. 1, pp. 275–276

Leino, P., Aro, S., Hasan, J. (1987). Trunk muscle function and low back disorders: a ten-year follow-up study. *Journal of Chronic Disorders*, Vol. 40, No. 4, pp. 289–296

Li, G., P. Buckle (1999). *Evaluating Change in Exposure to Risk for Musculoskeletal Disorders - a Practical Tool.* Available online at http://www.hse.gov.uk/research/crr_pdf/1999/crr99251.pdf. Suffolk, HSE Books CRR251

Mandel, M. A., (2003) Cumulative Trauma Disorders: History, Pathogenesis and treatment http://www.workinjuryhelp.com/ouch/mandel.htm, Accessed September 2011

Marras, W.S., Lavender, S.A., Leurgans, S.E., Fathallah, F.A., Ferguson, S.A., Allread WG, et al. (1995). Biomechanical risk factors for occupationally-related low back disorders. *Ergonomics*, Vol. *38*, No. 2, pp. 377–410

McAtamney, L., E. N. Corlett (1993). RULA: a survey method for the investigation of work-related upper limb disorders. *Applied Ergonomics*, Vol. 24, No. 2, pp. 91-99

McCauley Bush, P. (2011) *Ergonomics: Foundational Principles, Applications and Technologies*, an Ergonomics Textbook; CRC Press, Taylor & Francis, Boca Raton, FL

McCauley-Bell, P., Badiru, A. (1996a). Fuzzy Modeling and Analytic Hierarchy Processing to Quantify Risk Levels Associated with Occupational Injuries Part I: The Development of Fuzzy Linguistic Risk Levels. *IEEE Transactions on Fuzzy Systems*, Vol. 4, No. 2, pp. 124-131

McCauley-Bell, P., Badiru, A. (1996b). Fuzzy Modeling and Analytic Hierarchy Processing as a Means to Quantify Risk Levels Associated with Occupational Injuries Part II: The Development of a Fuzzy Rule-Based Model for the Prediction of Injury, *IEEE Transactions on Fuzzy Systems*, Vol. 4, No. 2, pp. 132-138

Moore, J. S., A. Garg (1995). The Strain Index: A proposed method to analyze jobs for risk of distal upper extremity disorders. *American Industrial Hygiene Association*, Vol. 56, pp. 443-458

Neese, R., Konz, S. (1993). Vibrometry of Industrial Workers: A Case Study. *International Journal of Industrial Ergonomics*, Vol. 11, No. 4, pp. 341-345

NIOSH (1991). Epidemiological basis for manual lifting guidelines. In: *Scientific support documentation for the revised 1991 NIOSH lifting equation: technical contract reports.* Springfield, VA: National Technical Information Service (NTIS), NTIS No. PB–91–226274.

NIOSH (1994). *Applications Manual for the Revised NIOSH Lifting Equation.* NIOSH Publication No. 94-110. (http://www.cdc.gov/niosh/docs/94-110/).

NIOSH (1997). *Musculoskeletal Disorders and Workplace Factors*, NIOSH Publication No. 97-141, http://www.cdc.gov/niosh/docs/97-141

NRC, Ed. (1999). *Work-Related Musculoskeletal Disorders*, National Research Council/National Academy Press

NRC, IOM (2001). *Musculoskeletal Disorders and the Workplace. Low Back and Upper Extremities*, National Research Council/Institute of Medicine//National Academy Press

Nunes, I. L., R. A. Ribeiro, et al. (1998). ERGO_X - A fuzzy expert system for the design of workstations. *Proceedings of the First World Congress on Ergonomics for Global Quality and Productivity (ERGON-AXIA '98)*, Hong Kong - China

Nunes, I. L. (2003). Modelo de Sistema Pericial Difuso para Apoio à Análise Ergonómica de Postos de Trabalho [Fuzzy Expert System Model to Support Workstation Ergonomic Analysis]. PhD thesis. Universidade Nova de Lisboa, Lisbon, Portugal: pp. 498 pages

Nunes, I. L. (2006). ERGO_X - The Model of a Fuzzy Expert System for Workstation Ergonomic Analysis. In: *International Encyclopedia of Ergonomics and Human Factors*, W. Karwowski (Ed). pp. 3114-3121. CRC Press, ISBN 041530430X

Nunes, I. L. (2009a). FAST ERGO_X - a tool for ergonomic auditing and work-related musculoskeletal disorders prevention. *WORK: A Journal of Prevention, Assessment, & Rehabilitation*, Vol. 34(2): 133-148

Nunes, I. L. (2009b). National Patent n. 103446 -Método de Análise Ergonómica de Postos de Trabalho [Workstation Ergonomic Analysis Method]

Occhipinti, E. (1998). OCRA: a concise index for the assessment of exposure to repetitive movements of the upper limbs. *Ergonomics*, Vol. 41, No. 9, pp. 1290-1311

Occhipinti, E., D. Colombini (2007). Updating reference values and predictive models of the OCRA method in the risk assessment of work-related musculoskeletal disorders of the upper limbs. *Ergonomics*, Vol. 50, No. 11, pp. 1727-1739

Okunribido, O., T. Wynn (2010). *Ageing and work-related musculoskeletal disorders. A review of the recent literature*, HSE.

Owen, B., Damron, C. (1984). Personal characteristics and back injury among hospital nursing personnel. *Res Nurs Health*, Vol. 7, pp. 305–313

Pransky, G., Hammer, K. Long, R., Schulz, L. Himmelstein, J., Fowke, J. (1997). Screening for Carpal Tunnel Syndrome in the Workplace: Analysis of Portable Nerve Conduction Devices. *Journal of Occupational and Environmental Medicine*, Vol. 39, No. 8, pp. 727-733

Putz-Anderson, V. (1988). *Cumulative Trauma Disorders: A Manual for Musculoskeletal Diseases of the Upper Limbs*, Taylor & Francis

Riihimäki, H., Tola, S., Videman, T., Hänninen, K. (1989). Low-back pain and occupation: a cross-sectional questionnaire study of men in machine operating, dynamic physical work, and sedentary work. *Spine*, Vol. 14, pp. 204–209

Ring, D., Kadzielski, J., Malhotra, L., Lee, S.-G. P., Jupiter, J. B. (2005). Psychological Factors Associated with Idiopathic Arm Pain. *The Journal of Bone and Joint Surgery (American)*, Vol. 87, No. 2, pp. 374-380

Sauter, S., Hales, T., Bernard, B., Fine, L, Petersen, M., Putz-Anderson, V., Schleiffer, L., Ochs, T. (1993). *Summary of two NIOSH field studies of musculoskeletal disorders and VDT work among telecommunications and newspaper workers*. In: Luczak, H., Cakir, A. & Cakir, G. (Eds.). Elsevier Science Publishers, B.V.

Svensson, H., Andersson G.B.J. (1983). Low-back pain in 40- to 47-year-old men: Work history and work environmental factors. *Spine*, Vol. 8, No. 3, pp. 272-276

Tanaka, S., McGlothlin D. J. (2001). A heuristic dose-response model for cumulative risk factors in repetitive manual work. In: *International Encyclopedia of Ergonomics and Human Factors*. W. Karwowski (Ed.), Taylor & Francis. III: 2646-2650

Terano, T., Asai, K., Sugeno, M. (1987). *Fuzzy Systems Theory and its Applications.* Academic Press, San Diego, CA

Tiilikainen, I. (1996). *WinOWAS - software for OWAS analysis. Tampere University of Technology,* Occupational Safety Engineering. (http://turva1.me.tut.fi/owas/).

Troup, J., Foreman, T., Baxter C., Brown, D. (1987). Musculoskeletal disorders at work, tests of manual working capacity and the prediction of low back pain. In: *Musculoskeletal disorders at work.* Buckle P. (Ed.) London, England: Taylor and Francis, pp. 165–170

Urbano, F.L. (2000). Tinel's sign and Phalen's maneuver: Physical signs of Carpal Tunnel syndrome. *Hospital Physician,* Vol. 36, No. 7, pp. 39–44

Vessey M.P., Villard-Mackintosh L. & Yeates D. (1990). Epidemiology of carpal tunnel syndrome in women of childbearing age. Findings in a large cohort study. *International Journal of Epidemiology,* Vol. 19, No. 3, pp. 655–659

Waters, T., Putz-Anderson, V., Garg, A. & Fine, L. (1993). 'Revised NIOSH equation for the design and evaluation of manual lifting tasks.' *Ergonomics,* Vol. 36, pp. 749-776.

Werne, R., Franzblau, A., & Johnston, E. (1994). Quantitative Vibrometry and Electrophysiological Assessment in Screening for Carpal Tunnel Syndrome Among Industrial Workers: A Comparison. *Archives of Physical Medicine and Rehabilitation,* Vol. 75, No. 11, pp. 1228-1232

Werner, S., Albers, J.W., Franzblau, A., Armstrong, T.J. (1994). The relationship between body mass index and the diagnosis of carpal tunnel syndrome. *Muscle & Nerve,* Vol. 17, No. 6, pp. 632-636

White, K., Congleton, J., Huchingson, R., Koppa, R., Pendleton, O. (1994). Vibrometry Testing for Carpal Tunnel Syndrome: a Longitudinal Study of Daily Variations. *Archives of Physical Medicine and Rehabilitation,* Vol. 75, No. 1, pp. 25-27

WHO (1985). Identification and control of work related diseases. Technical report series nº 714, World Health Organization.

Winn, F.J., Habes, D.J. (1990). Carpal tunnel area as risk factor for carpal tunnel syndrome. *Muscle Nerve,* Vol. 13, No 3, pp. 254–258

Women in America: Indicators of Social and Economic Well-Being (2011). Prepared by U. S. Department of Commerce Economics and Statistics Administration and Executive Office of the President, Office of Management and Budget

2

Work-Related Musculoskeletal Discomfort in the Shoulder due to Computer Use

Orhan Korhan
Department of Industrial Engineering, Eastern Mediterranean University, North Cyprus, Mersin
Turkey

1. Introduction

The National Institute for Occupational Safety and Health (NIOSH, 1997) in the USA defines Musculoskeletal Disorder (MSD) as a disorder that affects a part of the body's musculoskeletal system, which includes bones, nerves, tendons, ligaments, joints, cartilage, blood vessels and spinal discs. These are the injuries that result from repeated motions, vibrations and forces placed on human bodies while performing various job actions. The individual factors that can contribute to musculoskeletal symptoms include heredity, physical condition, previous injury, pregnancy, poor diet, and lifestyle.

Work-related musculoskeletal disorders occur when there is a mismatch between the physical requirements of the job and the physical capacity of the human body (Korhan, 2010). Musculoskeletal disorders are work-related when the work activities and work conditions significantly contribute to their occurrence, but not necessarily the sole or significant determinant of causation. Work-related musculoskeletal disorders (WRMSDs) describe a wide range of inflammatory and degenerative conditions affecting the muscles, tendons, ligaments, joint, peripheral nerves, and supporting blood vessels. These conditions result in pain and functional impairment and may affect especially the shoulder (Westgaard, 2000).

The causes of musculoskeletal disorders in the workplace are diverse and poorly understood. The meaning that working has to an individual may help to explain why certain psychological factors are associated with musculoskeletal discomfort and may eventually provide one way to intervene to reduce WRMSD (Mekhora et al., 2000) .

Musculoskeletal disorders have been observed and experienced widely at workplaces where the computers are frequently used. Increase in the number of employees working with computer and mouse coincides with an increase of work-related musculoskeletal disorders (WRMSDs) and sick leave, which affects the physical health of workers and pose financial burdens on the companies, governmental and non-governmental organizations (Korhan and Mackieh, 2010).

WRMSDs cover a wide range of inflammatory and degenerative diseases of the locomotor system, such as inflammations of tendons, pain and functional impairments of muscles, compression of nerves, and degenerative disorders occurring especially in the shoulder region due to occupations with large static work demands [European Agency for Safety and Health at Work (EU-OSHA), 2008].

The multifactorial causation of WRMSDs is commonly acknowledged. Different groups of risk factors including physical and mechanical factors, organizational and psychosocial factors, and individual and personal factors may contribute to the genesis of WRMSDs (EU-OSHA, 2008). Repetitive handling at high frequency, awkward and static postures, demanding and straining work and lack of recreation times, high time pressure, frequently overtime hours, repetitive or monotonous work, reduced physical capacity, obesity, and smoking are all the risk factors that contribute to WRMSDs either each one solely or by interacting each other.

WRMSDs largely affect the back (45%), and upper limb (37%); it is less common to suffer lower limb disorder (18%) (Health and Safety Executive, 2005). Work situations across all industries are implicated, particularly those involving use of the upper limbs, including computer work (Oakley, 2008).

This chapter presents the risk factors that contribute to musculoskeletal disorders in shoulders resulting from intensive use of computers in the workplaces. The risk factors of musculoskeletal disorders were revealed by assessing and analyzing workplace ergonomics, worker attitudes and experiences on the use of the computer keyboard and mouse. This was followed by an experimental data collection of muscle load, muscle force and muscular fatigue from the shoulder by Surface electromyogram (sEMG) to validate and verify the developed mathematical model.

Epidemiological studies in the literature confirmed that the work which is related with computer use brings higher risk for the development of musculoskeletal symptoms. Evans and Patterson (2000) tested the hypothesis that poor typing skill, hours of computer use, tension score and poor workstation setup are associated with neck and shoulder complaints, and they found out that tension score and gender were the only factors to predict neck and shoulder pain.

Jensen et al. (2002) found that the duration of computer work is associated with neck and shoulder symptoms in women, and hand symptoms in men. Additionally, the use of mouse was observed to have an increase in hand/wrist and shoulder region symptoms among the intensive users of computers.

Moreover, Karlqvist et al. (2002) concluded that for both genders the duration of computer work was associated with the musculoskeletal disorder symptoms, and women are at more risk of exposure to such disorder as they have less variability in work tasks.

Fogleman and Lewis (2002) studied the risk factors associated with the self-reported musculoskeletal discomfort in a population of video display terminal (VDT) operators, where their results indicated that there is a statistically significant increased risk of discomfort on each of the body regions (head and eyes, neck and upper back, lower back, shoulders, elbows and forearms, and hands and wrists) as the number of hour of keyboard use increases.

Blatter and Bongers (2002) studied the association of the effect of the gender differences with physical work factors as well as with the psychosocial factors. However their results showed that psychosocial factors were not related with the duration of computer use, whereas computer work of more than 6 hours per day was associated with musculoskeletal symptoms in all body regions of men, and computer work of more than 4 hours per day entailed the association with musculoskeletal disorders in women. Intensive computer use is associated with an increased risk of neck, shoulder, elbow, wrist and hand pain, paresthesias and numbness. Repetition, forceful exertions, awkward positions and localized contact stress are associated with the development of upper limb cumulative trauma in computer users.

Ming and Zaproudina (2003) showed that the repetitive computer use causes cumulative trauma on neck, shoulder, arm and hand muscles and joints.

In their model, Carayon et al. (1999) stipulated that psychosocial work factors (e.g. difficulty of job, working with deadlines, supervisor's pressure, lack of control), which can cause stress, might also influence or be related to ergonomic factors such as force, repetition, and posture that have been identified as risk factor for WRMSDs.

Peper et al. (2003) reviewed the ergonomic and psychosocial factors that affect musculoskeletal disorders at the workstation, and their results showed that there was a significant difference in right forearm extensor-flexor muscle tension and in right middle trapezius muscle tension between type tasks and rest.

Shuval and Donchin (2005) examined the relationship between ergonomic risk factors and upper extremity musculoskeletal symptoms in VDT workers, by taking into account individual and work organizational factors, and stress. Their results of RULA (Rapid Upper Limb Assessment) observations indicated that there were no acceptable postures of the employees whom were exposed to excessive postural loadings.

2. Methodology

2.1 Objectives

This research addresses worker perception and attitudes towards computer use, and their experiences with musculoskeletal symptoms in the shoulder and their diagnoses. The primary aim of this chapter is to present an in-debt assessment of the relationship between work-related musculoskeletal disorders in the shoulder and computer use. This study illustrates the idea of understanding how demographic structure (gender, age, height, and weight) physical and psychosocial job characteristics, office ergonomics, perceived musculoskeletal discomfort types and their frequencies may affect formation of musculoskeletal disorders in the shoulder. It then provides the evidence on the symptoms of musculoskeletal discomfort types and the frequency of these discomforts which are significant in the development of WRMSDs in the shoulder due to computer use.

The relevance of this study to the industry is to reduce the work-related musculoskeletal disorders associated with the intensive, repetitive and long period computer use that affect the shoulder. The developed risk assessment model also provides guidance for solving problems related to costly health problems (direct cost), lost productivity (indirect cost), and relieving the imposed economic burden.

As a summary, the research objectives of this study are:

- To assess and analyze workplace ergonomics, worker attitudes and experiences on computer use, and musculoskeletal symptoms in the shoulder developed by computer use,

- To determine a meaningful and statistically significant relationship between work-related musculoskeletal disorders in the shoulder and computer use, and develop a risk assessment model,
- To validate and verify the developed mathematical model through analysis of the data collected by the sEMG recordings.

2.2 Questionnaire

A questionnaire (see appendix) was developed based on the U.S. *National Institute for Occupational Safety and Health* (NIOSH) Symptoms Survey (NIOSH, 2011) and the Nordic Musculoskeletal Questionnaire (Dickinson et al., 1992). The questionnaire included questions in 7 modules according to the type of the questions. The questions were related with the demographic structure of the participant, physical job characteristics, psychosocial job characteristics, office ergonomics (workstation setup), types of musculoskeletal discomforts experienced at the shoulder, frequency of the musculoskeletal discomforts in the shoulder, and personal medical history.

The instrument was designed specifically for the current work. We are not aware of such an instrument being used for this purpose. In order to prevent any misunderstanding, the respondents were assisted at the time of answering the questionnaire.

2.3 Risk assessment model

In order to determine a meaningful and statistically significant relationship between work-related musculoskeletal disorders and computer use, a risk assessment model needs to be developed.

Logistic Regression Analysis was used to determine a meaningful and statistically significant relationship between shoulder discomfort and computer use, as a risk assessment model. The Logistic Regression was used since many of the independent variables were qualitative and the normality of residuals could not be guaranteed.

Our dependent variable was the WRMSD diagnosis made by a medical doctor (dichotomous dependent variable), and the independent variables were the rest of the variables in the questionnaire.

2.4 Experimentation

The respondents of the questionnaire, who have experienced musculoskeletal symptoms, were invited to a lab experiment, where surface electromyogram (sEMG) was used to record muscle load, muscle force and muscular fatigue. This test took place in two phases;

i. interrupting the work and performing test contractions of known force in a predetermined body posture and,
ii. comparing situations connected with a certain reference activity.

Before conducting the sEMG experiment, those respondents who were under high risk of having WRMSDs in the shoulder were identified using logistic regression. The significance level in logistic regression analysis was chosen to be 5% in order to minimize the possibility of making a Type I error. An independent variable with a regression coefficient not significantly different from 0 (p>0.05) can be removed from the regression model. If p<0.05 then the variable contributes significantly to the prediction of the outcome variable (Pampel, 2000).

Odd ratios of the significant factors for each respondent were calculated to find out respondents who were at the risk of having WRMSDs in the shoulder, as given below:

If χ_i's (i= 1,2,...) are independent variables, then the odds ratio is defined as

$$log\left[\frac{Prob\left(diagnosis\ of\ WRMSD\right)}{Prob\left(NOT\ diagnosis\ of\ WRMSD\right)}\right] = \beta_0 + \beta_1\chi_1 + \beta_2\chi_2 + ...$$

Where β_0 is the intercept, and $\beta_1, \beta_2, ...$ are the regression coefficients. Thus, we have $\text{Odds Ratio} = e^{(\beta_0 + \beta_1\chi_1 + \beta_2\chi_2 + ...)}$

In order to determine the respondents under high risk of having WRMSDs in the shoulder, the odds ratios of the significant factors for each respondent were calculated and those respondents who reflected maximum levels of odds ratios for each significant factor were invited for further investigation through electromyography.

Surface electromyogram was used to collect data from the shoulder. The procedure for the experimentation was as follows; twenty minutes typing exercise was given to each respondent at a time. Each respondent was asked to type a given standard text. Data were collected at 5th, 10th, 15th, and 20th minutes of the experiment. The mean value of the data collected for 30 seconds was then calculated and taken into consideration.

Analysis of Variance (ANOVA) and Factorial Analysis were applied at the end to the data collected by sEMG recordings, to validate and verify the significant risk factors of WRMSDs in the shoulder which were determined by logistic regression.

2.5 Respondents

A questionnaire was given to 130 people, who worked intensively with the computers for work/business purposes, such as; staff, research assistants and faculty members of Eastern Mediterranean University (EMU), web page designers, computer programmers, engineers, government officers, public relation officers, marketing officers, bank officers, customer representatives, commissioners, consultants, travel agents and translators. The reason for targeting such diverse disciplines was that the target population is expected to use computers intensively especially for work/business purposes and several other auxiliary purposes including personal and communication. Thus, the results were guaranteed not to be task-related, instead work-related.

3. Results

3.1 Descriptive statistics

Seventy male respondents (53.85%) attended this research. Males appeared to be dominating the female respondents (60, 46.15%). 107 (82.31%) of the 130 participants were between 20 to 35 years old.

40 respondents (30.77%) reported that their height were between 1.61-1.70 meters, which is followed by the height intervals 1.81-1.90 meters (35 respondents, 26.95%), and 1.71-1.80 meters (34 respondents, 26.15%).

The keyboard and mouse were reported to be the most popular (90.77%) input devices, whereas only 12 (9.23%) of the 130 respondents were using touchpad, keypad and trackball as primary input devices. Moreover, 88.43% of the respondents were using regular (Q-type) keyboards, 3.31% were using F-type keyboards, and 4.96% were using ergonomic (with wrist support) keyboards. Additionally, 72.31% of the respondents were using desktop and 27.69% of the respondents were using laptop computers.

Regarding the keyboard use, it was found that 55.04% of the respondents have been using keyboard for 10 or more years, and 37.98% have been using keyboard for at least 5 years.

Around 24.62% of the respondents reported their daily keyboard use as 5-6 hours per day, 23.85% of them as 7-8 hours per day, and 36.15% of them as more than 8 hours per day.

The results of the questionnaire indicated that 79.84% of the respondents found their job interesting, where 20.16% of the respondents indicated that they did not find their job interesting. Additionally, 74.42% of the respondents mentioned that their job gives them personal satisfaction; however 25.58% of the respondents mentioned that they were not having personal satisfaction from their job. A very high majority of the respondents (90.62%) reported that they have "good" relationship with their supervisor/advisor, where 9.38% reported that they have "not good" relationship with their supervisor/advisor.

More than two thirds of the respondents indicated that they share their office, where 35.66% share the office with more than three people, and 33.33% share the office with three or less people. On the other hand, 31.01% reported that they have their own office.

Majority of the respondents (84.38%) reported that they like their office environment, whereas 15.62% of the respondents reported that they do not like their office environment. Addition to this, a very high majority (94.57%) of the respondents indicated that they like working with computers, however only 5.43% of the respondents indicated that they do not like working with computers.

Most of the respondents (64.04%) reported that they have a stressful job, but 35.94% of the respondents reported that they do not have a stressful job. It was observed that, 48.84% of the respondents think that they have enough rest breaks, and 51.16% of the respondents do not think that they have enough rest breaks. Additionally, 46.88% of the respondents have repetitive (static) jobs, whereas 53.12% of the respondents have non-repetitive (dynamic) jobs.

Only 18.60% of the respondents were smokers when they answered the questionnaire, and 81.40% of the respondents were not smokers. More than half of the smoker respondents (63.41%) reported that they were smokers during the previous year, and 36.59% of the respondents were not smokers during the previous year.

Table 2 shows the results obtained on the workstation ergonomics. The results show that 56.92% of the respondents lean back to support their vertebrae, 67.69% reported that their feet were comfortable in the front, 81.54% stated that their seat and hands were centered on the keyboard, more than half (50.77%) of the respondents sit symmetrically, 79.23% use keyboard at the fingertips, 77.69% have the keyboard and the mouse at the same level, 80.62% of the respondents' screens were about arm length away from their eyes, 65.38% had the monitors at the eye level, 72.87% had sufficient lighting without glare, 78.46% had neutral wrist position, and 64.62% had neutral head and neck position.

However, the majority of the respondents didn't take into consideration of having 90^0 angle between the shoulders and the elbows. They did not care about sitting symmetrically at all, and they usually (73.64%) talked on the phone by having the handset between the head and the shoulder. Elbow, arm or leg supports also were not available in the respondents' workstations. Moreover, the majority of the respondents (78.46%) were not trained in posture (table 1).

Office Ergonomics	**Yes (%)**	**No (%)**
Lean back to support vertebrae	56.92	43.08
Elbows form 90 degrees flexion from shoulder	41.54	**58.46**
Feet are comfortable in the front of the chair	67.69	32.31
Seat and hands are centered on the keyboard	81.54	18.46
Sit symmetrically	50.77	49.23
Keyboard are at the fingertips	79.23	20.77
Keyboard and mouse are at the same level	77.69	22.31
Screen is arm length away from the eyes	80.62	19.38
Monitor is at the eye level	65.38	34.62
Sufficient lighting available, no glare	72.87	27.13
Talk on phone between head and shoulder	26.36	**73.64**
Neutral wrist position	78.46	21.54
Neutral head and neck position	64.62	35.38
Elbow and arm support available	48.84	**51.16**
Leg support available	25.58	**74.42**
Change sitting position every 15 min	57.69	42.21
Take active breaks	55.38	44.62
Take frequent microbreaks	45.38	**54.62**
Trained in posture	21.54	**78.46**

Table 1. Office ergonomics (n = 130).

Table 2 shows that the most prevalent discomfort experienced was having ache in the shoulder (46.15%). Discomfort (feeling of pain) was observed to be the next prevalent discomfort after ache. It was reported by the respondents that 34.62% of them were experiencing pain in the shoulder. Heaviness was reported by 17.69% of the respondents in the shoulder, and 9.23% of the respondents stated that they have a tightness in their shoulder. Having weakness was reported by 8.46% of the respondents in the shoulder, and having cramp in the shoulder was reported by 6.15%. Feeling of numbness was reported by 3.85% of the respondents and 3.08% of them reported tingling in their shoulder. Feeling hot and cold in the shoulder was reported by 2.31% of the respondents, and only 1.54% reported swelling in their shoulder.

	Percent Occurrence
Ache	46,15
Pain	34,62
Cramp	6,15
Tingling	3,08
Numbness	3,85
Heaviness	17,69
Weakness	8,46
Tightness	9,23
Feeling Hot and Cold	2,31
Swelling	1,54

Table 2. Type of discomfort and percent occurrence in the shoulder.

Therefore, the discomfort feelings of ache and pain were the most common types of discomforts which are experienced at the shoulder.

Table 3 shows the frequency of the discomforts experienced by the respondents.

	Never (%)	Rarely (%)	Sometimes (%)	Often (%)	Very Often (%)
Shoulder	8.46	10.77	26.92	17.69	12.31

Table 3. Frequency of discomfort.

Among the 130 respondents, 17 had a recent accident and 6 of those had this accident within 12 months (4.62% of the whole population). Also, 23 respondents reported that they had diagnosed with a work-related musculoskeletal disorder by a medical doctor, and 11 (8.46% of the whole respondents) of the sufferers reported this diagnosis had been made within the last 12 months.

Additionally, 4 respondents (3.08%) reported that they were diagnosed with rheumatoid arthritis, 1 respondent (0.77%) with diabetes, 4 respondents (3.08%) with thyroid disease, 8 respondents (6.15%) with pinched nerve. Moreover, 3 respondents were pregnant and 14 respondents with other medical symptoms and none of the respondents reported that they were diagnosed with hemophilia.

It was reported by the respondents that, 41 (31.54%) of them exercise never/rarely, 57 (43.85%) sometimes, 25 (19.23%) often, and only 7 (5.38%) of them exercise very often. Moreover, 91 of the respondents (70%) stated that they were involved in sport activities, and 39 of them (30%) reported that they were not involved in any kind of sport activities. More than half of the respondents (76, 58.46%) reported that they were involved in walking as sport activity, 17 of the respondents (13.08%) did jogging, 15 (11.54%) of them played football, 4 (3.08%) of them played basketball, 5 (3.85%) of them played volleyball, 10 (7.69%)

of them played tennis, 26 (20.00%) did swimming, and 27 (20.77%) involved in other sport activities.

3.2 Data analysis

Table 4 shows that only one of the above ergonomics factors, using keyboard and mouse at the same level (p=0.038<0.05) was found to be significant predictors of WRMSDs in the shoulder for the collected data.

Predictor	Coef	SE Coef	Z	P	Odds Ratio	95% CI Lower	95% CI Upper
Constant	7.02755	2.19910	3.20	0.001			
Elbow for 90^0	-0.720072	0.584639	-1.23	0.218	0.49	0.15	1.53
Sit symmetrically	0.0280327	0.594662	0.05	0.962	1.03	0.32	3.30
Centered hands	-0.513098	0.635498	-0.81	0.419	0.60	0.17	2.08
Monitor at eye level	-0.600702	0.524867	-1.14	0.252	0.20	0.55	1.53
Same level	-1.12705	0.544516	-2.07	0.038	0.11	0.32	0.94
Fingertips	-0.598543	0.542812	-1.10	0.270	0.55	0.19	1.59
Change sitting position	-0.421873	0.513102	-0.82	0.411	0.66	0.24	1.79
Elbow/arm support	0.271838	0.521752	0.52	0.602	1.31	0.47	3.65
Awkward tel use	-0.288873	0.600352	-0.48	0.630	0.75	0.23	2.43

Table 4. Logistic Regression of Ergonomic Factors that affects the Shoulder.

Table 5 shows that ache in the shoulder (p=0.024<0.05), pain in the shoulder (p=0.019<0.05), and having tightness in the shoulder (p=0.038<0.05) were found to be significant predictors of WRMSDs in the shoulder for the collected data.

Predictor	Coef	SE Coef	Z	P	Odds Ratio	95% CI Lower	95% CI Upper
Constant	2.16707	0.407966	5.31	0.000			
Ache	-0.473919	0.532888	-0.89	0.024	0.62	0.22	1.77
Pain	-0.673783	0.547809	-1.23	0.019	0.51	0.17	1.49
Cramp	0.137425	1.24328	0.11	0.912	1.15	0.10	13.12
Tingling	17.9883	14439.0	0.00	0.999	64896260.75	0.00	*
Numbness	19.9041	13118.6	0.00	0.999	4.40812E+08	0.00	*
Heaviness	-0.472714	0.631196	-0.75	0.454	0.62	0.18	2.15
Weakness	0.313680	1.04407	0.30	0.764	1.37	0.18	10.59
Tightness	-0.721157	0.752196	-0.96	0.038	0.49	0.11	2.12
Felling Hot&Cold	-0.609979	1.52017	-0.40	0.688	0.54	0.03	10.69
Swelling	-0.522971	1.66041	-0.31	0.753	0.59	0.02	15.35

Table 5. Logistic Regression of Feelings of Discomforts in the Shoulder.

Table 6 shows that often in the shoulder ($p=0.022<0.05$) was found to be significant predictors of WRMSDs for the collected data.

Predictor	Coef	SE Coef	Z	P	Odds Ratio	95% CI Lower	95% CI Upper
Constant	2.23359	0.607493	3.68	0.000			
Neck Never	-0.729515	0.990030	-0.74	0.461	0.48	0.07	3.36
Neck Rarely	-1.31730	0.847967	-1.55	0.120	0.27	0.05	1.41
Neck Sometimes	-0.185899	0.807035	-0.23	0.818	0.83	0.17	4.04
Neck Often	-1.19214	0.771063	-1.55	0.022	0.30	0.07	1.38
Neck Very Often	-1.13498	0.838082	-1.35	0.176	0.32	0.06	1.66

Table 6. Logistic Regression of Frequency of Discomforts in the Shoulder.

3.3 Experimental results

After developing the risk assessment model, the model should be validated and be verified. Towards this end, we have to first identify those respondents under risk. Then, the data analysis of the surface EMG recordings is supposed to provide the validation and verification.

Odds ratios for each significant factor determined by the logistic regression analysis were calculated and those respondents who have higher odds ratios for each factor were identified.

It was observed that fifteen respondents were under risk of having WRMSDs according to the results of odds ratio analysis. However, only six of the fifteen respondents were able to be contacted and invited to the sEMG data collection experiment. That group of six respondents formed the test group, and among the non-risk respondent group, six more respondents were invited to form the control group.

In the sEMG experiment, muscular activity in the shoulder (posterior deltoid) was recorded by using sEMG device (MyoTrac Infiniti, model SA9800). The procedure for the experimentation is as follows; 20 minutes typing exercise was given to each respondent at a time. Data were collected at 5th, 10th, 15th, and 20th minutes of the experimentation. The mean value of the collected data for 30 seconds is then calculated and taken into consideration.

3.3.1 Test group experimental results

The readings from sEMG provides the information about the muscle activity in the shoulder over time. Table 7 illustrates the mean value for each 30 seconds interval readings for each test group respondent.

The muscle activity is converted to μV by sEMG and is shown on the vertical axis, and time is shown on the horizontal axis in minutes (figure 1). Figure 1 illustrates that test group respondents have significantly high levels of muscle activities. There was a very significant

Muscle Activity Test Group	minutes			
	5	10	15	20
Respondent 1	319,8833	322,0783	333,4917	317,1383
Respondent 2	53,21833	51,12833	47,17	44,79333
Respondent 3	65,14167	277,8717	494,045	824,7967
Respondent 4	22,12667	21,44167	21,48833	23,85333
Respondent 5	510,13	346,92	571,84	232,0767
Respondent 6	135,7283	89,59	78,91833	53,97667

Table 7. Muscle activities (μV) of the test group respondents at the shoulder.

increase in the shoulder muscle activity of the test group respondent 3 throughout the experiment. Test group respondent 5 has been suffering from discomforts at the shoulder very significantly more than that of the other 5 respondents. Test group respondent 1 was experiencing almost a constant shoulder muscle activity during the experiment. The test group respondent 6 was observed to have a decreasing muscle activity during the experiment.

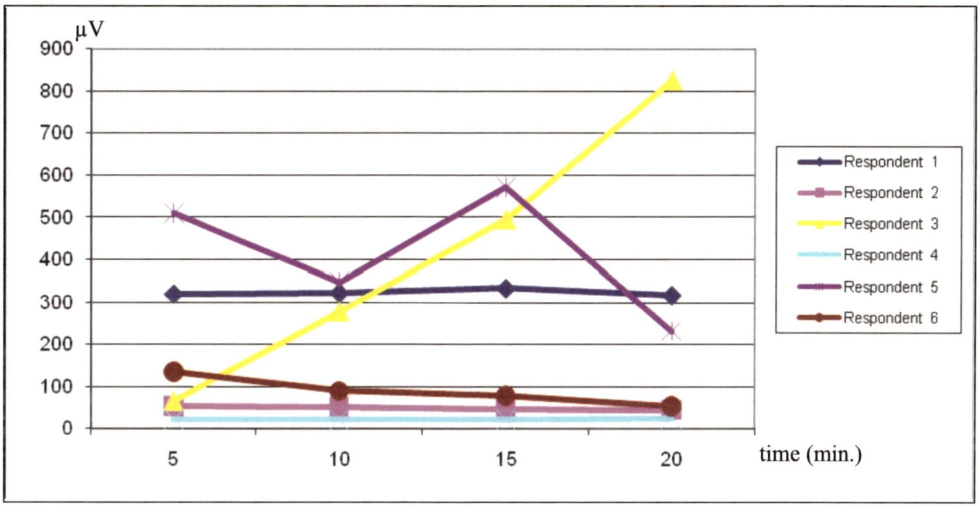

Fig. 1. Muscle activity recordings in the shoulder of test group respondents.

Table 8 shows the ANOVA results for the test group respondents.

Source of Variation	*SS*	*df*	*MS*	*F*	*P-value*	*F crit*
Between Groups	689220,1	5	137844	6,348314	0,001455	2,772853
Within Groups	390842,7	18	21713,48			
Total	1080063	23				

Table 8. ANOVA results for the test group respondents.

$F_0 = 6.348314 > F_{0.05,5,18} = 2.77$; therefore, reject . H_o .

Where;
H_0 = mean musculoskeletal strain (in time) of the 6 respondents does not differ, and
H_1 = mean musculoskeletal strain (in time) of the 6 respondents does not differs.

The results obtained by ANOVA indicate that, the risk assessment model developed has been validated and verified with the data collected through sEMG recordings.

3.3.2 Control group experimental results

The control group respondents were selected among the group of respondents who were not under risk according to the odds ratios.

Table 9 illustrates the mean value for each 30 seconds interval sEMG readings for each control group respondent.

Muscle Activity Control Group	minutes			
	5	10	15	20
Respondent 1	27,472	50,831	56,273	47,397
Respondent 2	19,11096	40,57785	62,46581	52,06078
Respondent 3	21,88277	58,33066	46,04155	81,68634
Respondent 4	25,374	33,002	176,6562	134,322
Respondent 5	22,978	89,7946	94,56764	162,2307
Respondent 6	19,69543	28,87675	28,62042	74,83737

Table 9. Muscle activities (μV) of the control group respondents at the shoulder.

Figure 2 illustrates that control group respondents' muscle activities do not significantly differ from each other and these readings were not at high levels. Moreover, the muscle activities of the control group respondents 4 and 5 showed slight but not significant increase throughout the experiment.

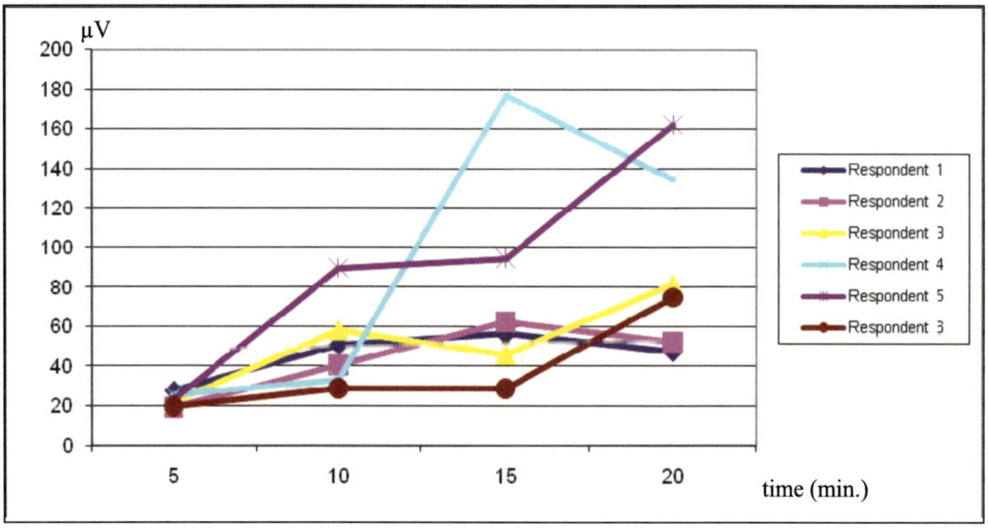

Fig. 2. Muscle activity recordings in the shoulder of control group respondents.

Table 10 shows the ANOVA results for the control group respondents.

Source of Variation	*SS*	*df*	*MS*	*F*	*P-value*	*F crit*
Between Groups	12486,11	5	2497,222	1,41259	0,266916	2,772853
Within Groups	31820,97	18	1767,832			
Total	44307,08	23				

Table 10. ANOVA results for the control group respondents.

$F_0 = 1.12594 < F_{0.05,5,18} = 2.77$; therefore, fail to reject. H_o ,

Where;
H_0 = mean musculoskeletal strain (in time) of the 6 respondents does not differ, and
H_1 = mean musculoskeletal strain (in time) of the 6 respondents does not differs.

The results of the ANOVA for each control group respondent indicate that, the mean musculoskeletal strain that they experience does not differ in time. That is, the musculoskeletal strain at their shoulder do not differ as those in the test group respondents.

ANOVA results for the control group respondents support the risk assessment model developed to determine the risk factors of WRMSDs.

4. Conclusion

Most of the studies on the formation of WRMSDs during computer use have been focused on the gender differences, physical and psychological aspects of the user and have not yet considered extra-rational factors such as the perceived musculoskeletal discomfort types and their frequencies. This study presents the idea of understanding how office ergonomics, perceived musculoskeletal discomfort types and their frequencies may affect formation of musculoskeletal disorders at the shoulder.

After collecting data from 130 respondents, the significant findings related with discomfort in shoulder during computer use were:

- Using keyboard and mouse at the same level [OR=0.11, CI: 0.32-0.94]
- Ache in the shoulder [OR=0.62, CI: 0.22-1.77]
- Pain in the shoulder [OR=0.51, CI: 0.17-1.49]
- Having tightness in the shoulder [OR=0.49, CI: 0.11-2.12]
- Often in the neck [OR=0.30, CI: 0.07-1.38]

This study provided the evidence that, for the study groups tested and for the given computer use activity, ache and pain are the most common types of the discomforts in the shoulder. Also, this study showed that the mean musculoskeletal strain at the shoulder of test group respondents differ in time, whereas for each control group respondent, the mean musculoskeletal strain that they experience, does not differ in time.

5. Appendix: Questionnaire

Name, Surname: Occupation:

Tel no: E-mail:

1. What is your gender?

☐ Male ☐ Female

2. What is your age?

☐ 20-25

☐ 26-30

☐ 31-35

☐ 36-40

☐ 41-45

☐ 46-50

☐ Older than 50

3. How tall are you in meters?

☐ Shorter than or equal to 1.50

☐ 1.51-1.60

☐ 1.61-1.70

☐ 1.71-1.80

☐ 1.81-1.90

☐ 1.91-2.00

☐ Taller than 2.00

4. How much do you weigh in kilograms?

☐ Less than or equal to 50

☐ 51-60

☐ 61-70

☐ 71-80

☐ 81-90

☐ 91-100

☐ More than 100

5. What type of *computer* do you mostly use? (mark only one)

☐ Desktop

☐ Laptop

☐ PDA (Personal Digital Assistant) / Pocket PC

☐ Mainframe

☐ Minicomputer

☐ Server

6. What type of computer input devices do you mostly use? (mark only one)

☐ Keyboard and mouse

☐ Touchpad and keypad

☐ Trackball

☐ Touch pen

☐ Joystick and joypad

7. Typically, how much time daily in total you spend typing on a computer keyboard or using a mouse?

☐ Less than 1 hour

☐ 1-2 hours

☐ 3-4 hours

☐ 5-6 hours

☐ 7-8 hours

☐ More than 8 hours

8. Overall, how many years have you been using computers?

☐ Less than 1 year

☐ 1-2 years

☐ 3-4 years

☐ 5-9 years

☐ More than 10 years

9. What type of computer *keyboard* you mostly use? (mark only one)

☐ Regular (Q-type)

☐ Regular (F-type)

☐ Ergonomic (with wrist support)

☐ Other (Please specify)

10. Do you think you have an interesting job?

☐ Yes ☐ No

11. Does your current job give you *personal* satisfaction?

☐ Yes ☐ No

12. How do you define your relationship with your current supervisor/advisor?

☐ Good☐ Not good

13. In what kind of office environment you work?

☐ I share the office with more than 3 people

☐ I share the office with 3 or less people

☐ I have my own office

14. Do you like your office environment?

☐ Yes ☐ No

15. Do you like working with computers?

☐ Yes ☐ No

16. Do you think you have a stressful job?

☐ Yes ☐ No

17. What kind of job you have?

☐ Repetitive (Static) ☐ Non-repetitive (Dynamic)

18. Do you think you have enough rest breaks?

☐ Yes ☐ No

19. Do you currently smoke?

☐ Yes ☐ No

If "Yes", have you been smoking in the last year?

☐ Yes ☐ No

The questions in the table below are related with your working posture. Mark with "Yes" if the statement is applicable, mark with "No", if it is not applicable with your working posture.

	YES	NO
20. Lean back in chair to support your vertebrae	☐	☐
21. Elbows form a 90 degree angle while hanging at sides from the shoulders	☐	☐
22. Feet are comfortable on the floor in front of you	☐	☐
23. Your seat and your hands are centered on the keyboard	☐	☐
24. Sit symmetrically (not bending either sides)	☐	☐
25. The keyboard and the mouse are at the fingertips	☐	☐
26. The keyboard and the mouse are on the same level (side by side)	☐	☐
27. The screen is about an arm's length away from the eyes	☐	☐
28. The top of the monitor is at the eye level	☐	☐
29. Sufficient lightening available without glare from lights, windows, surfaces, and etc…	☐	☐
30. Frequent use of telephone between head and shoulder	☐	☐
31. Neutral position of the wrist (straight from fingers to the elbow)	☐	☐
32. Neutral position of the head and the neck	☐	☐
33. Elbow/arm support provided for intensive/long durations	☐	☐
34. Leg support provided for intensive/long durations	☐	☐
35. Change sitting position at least every 15 minutes	☐	☐
36. Take active breaks (phone call, file paper, drink water, etc…) every 30 minutes	☐	☐
37. Take frequent microbreaks (while seated on your workstation)	☐	☐
38. Trained in proper posture	☐	☐

39. During the last 12 months, have you experienced, *while using a keyboard or a mouse,* the following symptoms in the following body regions? (mark all that apply)

	Neck	Shoulder	Elbow/ forearm	Hand/ wrist	Finger	Upper back	Lower back
Ache	☐	☐	☐	☐	☐	☐	☐
Pain	☐	☐	☐	☐	☐	☐	☐
Cramp	☐	☐	☐	☐	☐	☐	☐
Tingling	☐	☐	☐	☐	☐	☐	☐
Numbness	☐	☐	☐	☐	☐	☐	☐
Heaviness	☐	☐	☐	☐	☐	☐	☐
Weakness	☐	☐	☐	☐	☐	☐	☐
Tightness	☐	☐	☐	☐	☐	☐	☐
Feeling Hot & Cold	☐	☐	☐	☐	☐	☐	☐
Swelling	☐	☐	☐	☐	☐	☐	☐

40. How often have you experienced those symptoms? (mark all that apply)

	Never	Rarely	Sometimes	Often	Very Often
Neck	☐	☐	☐	☐	☐
Shoulder	☐	☐	☐	☐	☐
Elbow/forearm	☐	☐	☐	☐	☐
Hand/wrist	☐	☐	☐	☐	☐
Finger	☐	☐	☐	☐	☐
Upper back	☐	☐	☐	☐	☐
Lower back	☐	☐	☐	☐	☐

41. Have you had any recent accident?

☐ Yes ☐ No

If "Yes", when?

☐ Within 1 year

☐ More than 1 year

42. Have you been diagnosed with any of the following medical symptoms? (mark all that apply)

☐ Yes ☐ No

If "Yes", which one(s)?

☐ Rheumatoid *arthritis*

☐ Diabetes

☐ Tyroid disease

☐ Hemophilia

☐ Pinched nerve

☐ Recent pregnancy

☐ Other, please specify

43. Have you been diagnosed by a medical doctor with work-related musculoskeletal disorders (herniated disk, carpal tunnel syndrome, tendonitis, etc…)?

☐ Yes ☐ No

If "Yes", have you been diagnosed within 12 months?

☐ Yes

☐ No

If "No", has a medical doctor ever told you that you are at risk for work-related musculoskeletal disorders?

☐ Yes ☐ No

44. Do you exercise?

☐ Never or rarely ☐ Sometimes ☐ Often ☐ Very often or constantly

45. Are you involved in any of the following sport activities?

☐ Yes ☐ No

If "Yes", which one(s)? (mark all that apply)

☐ Walking ☐ Football ☐ Basketball ☐ Swimming

☐ Jogging ☐ Volleyball ☐ Tennis Other, please specify

6. References

Blatter, B. M., Bongers, P. M. (2002). Duration of computer use and mouse use in relation to musculoskeletal disorders of neck or upper limb. *International Journal of Industrial Ergonomics*, Vol. 30, 295-306, ISSN 0169-8141.

Carayon, P., Smith, M. J., Haims, M. C. (1999). "Work Organization, Job Stress, and Work-Related Musculoskeletal Disorders", *Human Factors*, Vol. 41 (4): 644-663, ISSN 1520-6564.

Dickinson, C.E., Campion, K., Foster, A.F., Newman, S.J., O'Rourke, A.M., Thomas, P.G. (1992). Questionnaire development: an examination of the Nordic Musculoskeletal questionnaire. *Applied Ergonomics* 23(3): 197-201, ISSN 0003-6870.

Evans, O., Patterson, K., (2000). "Predictors of neck and shoulder pain in non-secretarial computer users", *International Journal of Industrial Ergonomics*, Vol. 26: 357-365, ISSN 0169-8141.

EU-OSHA (2008). What are work-related musculoskeletal disorders (WRMSDs)? In: *European Agency for Safety and Health at Work,* 2012, Available from: http://osha.europa.eu/en/faq/frequently-asked-questions/what-are-work-related-musculoskeletal-disorders-msds.

Fogleman M., Lewis, R. J. (2002). Factors associated with self-reported musculoskeletal discomfort in video display terminal (VDT) users. *International Journal of Industrial Ergonomics,* Vol. 29, 311-318, ISSN 0169-8141.

Health and Safety Executive (HSE, 2005). Health and Safety Statistics 2004/05. In: Health and Safety Commission, 2012. Available from: http://www.hse.gov.uk/statistics/overall/hssh0405.pdf

Jensen, C., Finsen, L., Søgaard, K., Christensen, H. (2002). Musculoskeletal symptoms and duration of computer and mouse. *International Journal of Industrial Ergonomics,* Vol. 30 (4-5), 265-275, ISSN 0169-8141.

Karlqvist, L., Tornqvist, E. W., Magberg, M., Hagman, M., Toomingas, A. (2002). Self-reported working conditions of VDU operators and associations with musculoskeletal symptoms: a cross-sectional study focusing on gender differences. *International Journal of Industrial Ergonomics,* Vol. 30, 277-294, ISSN 0169-8141.

Korhan, Orhan. "Work-Related Musculoskeletal Disorders due to Computer Use: A Literature Review". Lambert Academic Publishing, Germany 2010. ISBN: 978-3-8383-6625-8.

Korhan, O., Mackieh A. (2010). A Model for Occupational Injury Risk Assessment of Musculoskeletal Discomfort and Their Frequencies in Computer Users. *Safety Science,* Vol. 48 (7), 868-877, ISSN 0925-7535.

Mekhora K., Liston, C. B., Nanthanvanij, S., Cole, J. H. (2000). The effect of ergonomic intervention on discomfort in computer users with tension neck syndrome. *International Journal of Industrial Ergonomics,* Vol. 26, 367,-379, ISSN 0169-8141.

Ming, Z., Zaproudina, N. (2003). Computer use related upper limb musculoskeletal (ComRULM) disorders. *Pathophysiology,* Vol. 9, 155-160, ISSN 0928-4680.

NIOSH, "Elements of Ergonomics Programs: A Primer Based on Evaluations of Musculoskeletal Disorders," 1997 DHHS (NIOSH) Publication No. 97-117, ISSN 0348-2138.

NIOSH (2011). Elements of Ergonomics Programs. In: *NIOSH Publication No. 97-117,* 2012, Available from: http://www.cdc.gov/niosh/docs/97-117/eptbtr4.html

Oakley, Katie. "Occupational Health Nursing". John Wiley & Sons Inc, USA 2008. ISBN: 9780470035335.

Pampel, Fred C. "Logistic Regression, A Primer". SAGE Publications Inc., USA 2000. ISBN: 9780761920106.

Peper, E., Wilson, V. S., Gibney, K. H., Huber, K., Harvey, R., Shumay, D.,M. (2003). "The integration of electromyography (SEMG) at the workstation: assessment, treatment, and prevention of repetitive stain injury (RSI)", *Applied Psychophysiology and Biofeedback,* Vol. 28 (2): 167-182, ISSN 1573-3270.

Shuval, K., Donchin, M. (2005). "Prevalence of upper extremity musculoskeletal symptoms and ergonomic risk factors at a High-Tech company in Israel", *International Journal of Industrial Ergonomics,* Vol. 35: 569-581, ISSN 0169-8141.

Westgaard, R. H. (2000). Work-related musculoskeletal complaints: some ergonomics challenges upon the start of a new century. *Applied* Ergonomics, Vol. 31, 569-580, ISSN 0003-6870.

3

Ergonomic Impact of Spinal Loading and Recovery Positions on Intervertebral Disc Health: Strategies for Prevention and Management of Low Back Pain

S. Christopher Owens[1],
Dale A. Gerke[2] and Jean-Michel Brismée[3]
[1]*Hampton University Doctor of Physical Therapy Program, Hampton, VA,*
[2]*Concordia University Wisconsin, Mequon, WI,*
[3]*Center for Rehabilitation Research, Clinical Musculoskeletal Research Laboratory, Texas Tech University Health Sciences Center, Lubbock, TX, USA*

1. Introduction

According to the United States Department of Labor, over three million nonfatal injuries and illnesses occurred in private industry during the year 2009. Of these work related injuries 195,150 injuries involved the lumbar spine (U.S. Department of Labor, Bureau of Labor Statistics, 2009). The cost of work related low back injuries has been estimated to exceed 16 billion dollars. Low back pain accounts for an estimated 149 million lost work days per year. The estimated cost of this lost productivity is $28 billion (U.S. Department of Labor, Bureau of Labor Statistics, 2009). The costs in terms of medical care as well as lost productivity have brought the prevention of these injuries to the forefront in occupational medicine.

Management of low back pain, particularly work related injuries, is very controversial with numerous different treatment approaches ranging from osteopathic manipulations to work hardening programs. However, these strategies have been marked by many non-scientific interventions. A comprehensive understanding is essential for clinicians to implement effective evidence-based treatment and prevention. The purposes of this chapter are to (1) review the anatomical, biomechanical, and physiological mechanisms that contribute to the health of the lumbar spine with particular emphasis on the IVD, (2) consider mechanisms that may cause pain and dysfunction in the lumbar spine, and (3) present specific strategies for prevention and management of work related low back pain based on the biomechanical and physiological response of the lumbar IVD.

2. Intervertebral disc anatomy and physiology

The functional spinal unit consists of the two adjacent vertebral bodies, the IVD, and the adjoining ligaments and fascia that cross the segment. In comparison to the axial spine as a whole, the lumbar IVD to vertebral body ratio is the largest in the lumbar spine, approximately 1:3 ratio. Along with action of the iliopsoas muscle, it is the intervertebral disc height that accounts for the normal lordotic posture observed in the sagittal plane. The intervertebral disc provides resistance to compressive loads at the spine while simultaneously allowing very complex multi-planar movements to occur (Urban & Roberts, 2003). The intervertebral disc can be divided into three separate regions; anulus fibrosus, nucleus pulposus, and cartilaginous endplate (Sizer et al., 2001).

The annulus fibrosus forms the outer walls of the IVD. The annulus is considered fibrous cartilage and is comprised of predominantly type I collagen (Urban & Roberts, 2003). The fibers of the annulus are arranged in fifteen to twenty five concentric lamellae at approximately 60 degree angles from the vertical plane (Urban & Roberts, 2003). The annulus has three distinct zones. The outer third attached to the outer aspect of the adjacent vertebral bodies. The middle third directly attaches to the cortex of the vertebral body, while the inner third is confluent with the cartilaginous end-plate and creates a continuous envelope around the nucleus pulposus (Sizer et al. 2001). Innervation is only present in the outer third of the annulus in healthy intervertebral discs (Urban & Roberts, 2003). The posterior portion of the annulus is innervated by the sinuvertebral nerve while the anterior portion of the outer annulus receives sensory fibers through the paravertebral sympathetic trunks (Morinaga et al., 1996).

The nucleus pulposus is a gelatinous, highly hydrated structure comprised primarily of type II collagen and water binding proteoglycans. It is sandwiched between the cartilaginous endplate inferiorly and superiorly. The endplate is comprised of hyaline cartilage, and is the boundary between the vertebral body and the IVD.

2.1 Intervertebral disc nutrition

The intervertebral disc is primarily an avascular structure with the reported occurrence of blood vessels in the outer most anulus fibrosus being very rare (Crock & Goldwasser, 1984; Freemont et al., 1997). Without the presence of blood vessels the IVD is primarily dependent on diffusion for small solutes and fluid flow for larger protein molecules (McMillan et al., 1996). Nutrient diffusion into the IVD is linked to the cycle of lost water with compression and the in-flow of water with removal of loads (Sehgal & Fortin, 2000).

The intervertebral disc consists of a relatively small concentration of fibroblasts, which are the cells responsible for production of collagen and the proteins that comprise the extracellular matrix (Nordin & Frankel, 1989). All cells including fibroblasts utilize glucose and oxygen for energy, and produce waste products such as lactate. Selard and colleagues found that concentrations of oxygen and glucose, within the lumbar IVD, were lowest in the center of the nucleus pulposus (Selard et al., 2003). This was also the area where the highest level of lactate was found. This is presumably a result of the central nucleus being at the furthest point from the endplate vertebral body interface. These small oxygen and glucose molecules are primarily supplied to the nucleus pulposus via diffusion through the cartilaginous end plate (Holm et al., 1981; Rajasekaran et al., 2004). Rajasekaran and colleagues confirmed the primary role of the cartilaginous endplate with their in-vivo magnetic resonance imaging (MRI) study that documented the diffusion patterns of injected gadodiamide (Rajasekaran et al., 2004). Degeneration is a part of the normal aging process that all tissues experience, while IVD degradation is an acceleration of this process, and occurs under pathological conditions where the normal balance between nutrition and waste elimination fails. The delineation between degradation and normal age related degeneration is very difficult to categorize.

2.2 Normal aging in the lumbar intervertebral disc

The normal degenerative process that occurs in the IVD begins in the second decade of life when endplate vascularity gradually decreases to a complete absence of vascular tissue (Boos et al., 2002). This is a progressive process with peak degenerative alterations normally occurring between fifty and seventy years of age, with the degenerative process being largely influenced by genetic factors (Battié et al., 2009; Kalichman & Hunter, 2008). Several authors have reported altered IVD nutrition as being the primary catalyst for the age related degenerative process (Boos et al., 2002; Rajasekaran et al., 2004; Urban & McMullin, 1988). The loss of endplate vascularity beginning in the second decade substantially decreases diffusion of vital nutrients to the nucleus.

Degenerative changes at the cartilaginous end plate precede changes within the nucleus (Boos et al., 2002). These histological changes include disorganization of collagen fibers and mucoid degeneration. In addition, sclerosis of the vertebral body can also occur in advanced stages cartilaginous end plate degeneration. As previously stated, the cartilaginous endplate plays a primary role in the nutritional status of the IVD, and disruption of the endplate hastens the degenerative process. It has been suggested that a classification system based on endplate diffusion characteristics may be the most appropriate way of determining the differences between normal aging and pathological degradation (Rajasekaran et al., 2004).

The nucleus pulposus undergoes histological changes with degeneration as well. The nucleus becomes more fibrotic with increased disorganization of collagen fibers. The most critical change that occurs is a decrease in proteoglycan content (Urban & Roberts, 2003). Proteoglycans bind water and maintain the normal hydration levels of the nucleus. This decrease in proteoglycan content is reflected in the decreased hydration levels seen in degenerated IVDs (Urban & McMullin, 1988).

The morphological changes that occur in the annulus fibrosus during IVD degeneration include: radial fissures within the annulus, disassociation of lamellar fibers with resulting bulging (inward inner third and outward outer third), and mucoid degeneration (Adams et al., 2000). Another consequence of degeneration with important clinical implications is nerve in-growth into the inner two thirds of the annulus (Coppes et al., 1997). This further increases the possibility of pain generation from the IVD.

3. Lumbar intervertebral discogenic pain

The degenerative histological changes that occur in the annulus fibrosus are most frequently associated with discogenic pain (Adams et al., 1996; Edwards et al., 2001; Zhao et al., 2005). The results of proteoglycan content and decreased hydration within the nucleus pulposus is diminished load distribution (Buckwalter, 1995). Zhao and colleagues proposed IVD disc dehydration, and the resulting loss in segmental height, as being one of the possible causes of pain in degenerated discs (Zhao et al., 2005). They hypothesized that pain in dehydrated degenerative discs occurred as a result of increased stress concentrations within the nociceptive posterior annulus. Adams and colleagues found similar high stress peaks in the posterior annulus of degenerated IVD in their study of in-vitro IVD stress profiles (Adams et al., 1996). McNally and colleagues assessed in-vivo lumbar IVD disc stress profiles in patients undergoing provocative discography, and found that pain was predictive of abnormal posterior annulus stress concentrations and depressurized nucleus pulposus (McNally et al., 1996). These findings suggest that the loss of hydration and the loss of segmental height associated with lumbar IVD degeneration may be an important contributing factor in mechanical low back pain.

Although the origins of low back pain are widely debated and imaging has done little to clarify this debate, the IVD is the most common source of low back pain in adults (DePalma et al., 2011). There are imaging findings that are characteristic of IVD degeneration; however, these findings do not necessarily correlate with the symptoms associated with low back pain. Based on MRI findings, lumbar IVD herniation rates in asymptomatic populations have been found to be as high as 76%, (Boos et al., 1995). Deyo and colleagues estimated that 85% of individuals with low back pain may experience non-specific low back pain (Deyo et al., 1996). This indicates that the vast majority of individuals experiencing low back pain will not have any specific diagnostic imaging findings that explain their symptoms.

The Magnetic Resonance Imaging finding most associated with IVD degeneration is diminished nucleus signal intensity on T2 weighted images. In advanced stages of degeneration, narrowing of the IVD space can be observed on plane radiographs. High-intensity zones within the annulus fibrosus on T2 weighted MRI are associated with degenerative annular tears (Schmidt et al., 1998).

With the absence of definitive imaging to identify symptom related IVD degeneration, subjective and objective clinical findings can provide important information on the functional spinal unit. Cook and colleagues in a Delphi study of physical therapists reported common subjective and objective signs of non objectifiable instability (Cook et al., 2006). One of the consequences of IVD degeneration and diminished hydration is a reduction in segmental stability. Zhao and clleagues proposed IVD dehydration and resulting increases in stress concentrations in the posterior annulus as being one of the possible causes of pain in degenerated discs (Zhao et al., 2005). The subjective and objective findings associated with this diminished stability include: long history of intermittent back problems, complaints of "catching" sensations, transient neurological symptoms/deficits, experience of "twinges", minor activities causing significant complaints, rotation causing sharp shooting pain, and pain with prolonged activities (Cook et al., 2006).

Loss in IVD height and hydration can result in decreased mechanical energy dissipation, radial IVD bulging, and increased zygapophyseal joint loading (Adams et al., 1990) Cinotti and colleagues reported diminished foraminal height with narrowing of the IVD space (Cinotti et al., 2002). They hypothesized that this loss in foraminal height, along with resulting buckling of the ligamentum flavum, may be the cause of the symptoms associated with foraminal stenosis.

4. Lumbar intervertebral disc loading and recovery postures

Throughout the course of the day, the Lumbar IVD demonstrate viscoelastic creep properties that determine the overall stature of an individual. Tyrrell and colleagues used in-vivo stadiometry measurements to detect 19.3 mm (1.1% of stature) variation in height between first arising and the end of the day (Tyrrell et al., 1985). Paajanen and colleagues using MRI to confirm the role of the intervertebral disc, reported similar results with subjects losing 13 and 21 mm of height during the day (Paajanen et al., 1994). Stadiometry and MRI are the two primary methods of measuring spinal height change following loading and recovery conditions. Stadiometry has been shown to be a valid and reliable clinical tool to assess spinal height when compared to quantifiable measures made from MRI (Kanlayanaphotporn et al., 2002; Kourtis et al., 2004; Owens et al., 2009; Pennell et al., 2012). Stadiometry assessment has advantages over MRI in terms of costs, use in clinical setting, as well as the ability to measure subjects that simultaneously sustain compressive loads of the trunk.

Several authors have assessed the ergonomic impact of work related spinal loading. Eklund and Corlett used stadiometry to assess the specific effects of work related postures and activities including types of office chairs and standing activities (Eklund & Corlett, 1984). Helander and colleagues used a stadiometer to compare changes in height following periods of prolonged sitting that were accompanied by either standing or walking rest breaks (Helander et al., 1990). Leivseth and Drerup, also measured spine height, to assess the impact of sustained sitting and standing work activities with greatest shrinkage occurring during work activities in standing (Leivseth & Drerup, 1997).

Static loading, particularly while sitting, has been associated with increased work related low back pain (Fryer et al., 2010). Knowledge of interventions and postures that can potentially offset these affects can have an important impact on treatment and prevention of work related low back pain. A primary focus of ergonomics research been on recovery positions designed to restore spinal height. Magnusson and colleagues reported greater height recovery following loaded sitting, with ten minutes of prone hyperextension lying when compared to prone lying in neutral (Magnusson et al. 1996). Additional studies have demonstrated that sustained supine flexion and sidelying flexion position also increase spine height following periods of seated loading. (Gerke et al., 2011; Owens et al., 2009).

5. Prevention and management strategies for low back pain in the workplace

Management of low back pain secondary to disc related disorders can be challenging for the patient and clinician. Providing appropriate ergonomic suggestions based on the biomechanics of the lumbar IVD can improve the tolerance to work. Ergonomic suggestions that aim to maintain an optimal amount of disc hydration while minimizing disc pressure will be discussed.

The sitting position is a common quandary for individuals experiencing back pain secondary to a disc related disorder. Sitting is generally not very well tolerated by an individual. However, the sitting position is difficult to avoid during travel to work. Sitting is also a common position adopted at work. Therefore, the sitting position should be carefully evaluated if there are discogenic symptoms.

Sitting position without support to the lumbar spine creates nearly 50 % increase in IVD pressure compared to sitting with lumbar support. Previous research by Wilke and colleagues demonstrated how various sitting positions can affect the pressure of the IVD (Wilke et al. 2001). Selecting a chair with back support is ideal for the person experiencing low back pain related to IVD pathology. In contrast, sitting on a stool or chair without adequate thoracolumbar back support can cause an increase in disc pressure as previously measured (Wilke et al., 2001). In many work settings, a specific back support may be suggested by a clinician. Location of the back support is often identified as a position of greatest comfort for the individual. A back rest in the thoracic spine may also be recommended if the goal is to bring more surface contact to the spine. As suggested by previous research, thoracic spine support decreased the amount of, thoracic spine support decreased the amount of intra-discal pressure in the lumbar spine (Wilke et al., 2001). In addition, it is important for individuals to have adequate foot support if the occupation requires a large amount of time in the sitting position. Proper foot support provides stability to the spine and decreases use of the abdominal muscles while in sitting. If the feet are unsupported, the weight of the legs can increase the lordosis creating an uneven stress distribution in the lumbar spine. In addition, tension from a tight iliopsoas muscle can also create the potential for more lordosis if the feet are unsupported in sitting.

When working with patients experiencing discogenic low back pain, it is also important for clinicians to consider the time of day as it relates to IVD hydration. Sleeping for greater than 6 hours will allow the IVD to imbibe fluid. As indicated previously, the amount of IVD hydration following a prolonged unloaded position such as sleeping can cause an increase in IVD pressure immediately after waking. The patient may feel stiffness in the lumbar region or difficulty standing up straight. After moving the trunk or walking for several minutes to a few hours, the stiffness may subside, allowing more freedom with movement. A creep response allows the disc to dehydrate. The IVD dehydration that occurs with moving and walking will decrease lumbar intradiscal pressure following a period of rest in supine or sidelying position. Therefore, a person may feel more comfortable with lumbar motion following activities that dehydrate the lumbar spine rather than immediately after waking. Lumbar range of motion may also increase after the lumbar intervertebral disc has been cautiously loaded for a brief period of time secondary to dehydrating the lumbar disc. Based on the increased hydration and increased IVD pressure associated with first arising in the morning, individuals with lumbar discogenic pain should avoid forward bending immediately after waking (Snook et al., 1998). Time should be allowed for the lumbar IVD pressure to decrease as a result of normal loading. In fact, it is advisable for the patient experiencing low back pain secondary to a lumbar IVD related disorder to be active in an upright position immediately after waking. In addition, the clinician may suggest gentle, pain free, repetitive lumbar range of motion to assist with dehydrating the lumbar disc before performing forward bending activities. As the day progresses, it may be safer to perform activities that involve a larger range of motion including flexion exercises (Table 1).

Lifting can also increase pressure in the IVD. If the integrity of the posterior annular fibers of the lumbar IVD's are compromised, lifting will increase the load placed on the IVD. Therefore, heavy lifting soon after waking should be avoided. Lifting with loads close to the body will create a shorter external moment arm for the lumbar musculature. Lifting with loads further from the body creates a large external moment arm and subsequently requires the thoracolumbar extensors to contract with greater force. The increase in contraction by these thoracolumbar extensors will create an increase in IVD pressure to counter the external moment produced by the load anterior to the body. Hence, body mechanics instructions incorporate advice to keep external loads close to the body. Even small loads may produce a large increase in IVD pressure.

	Intervertebral Lumbar Disc Dehydration Activities	**Intervertebral Lumbar Disc Re-hydration Activities**
Types of activities	- Gravity activities Walking Jogging - Repetitive cyclic trunk motions (Figure 4 - 5) - Trunk & pelvic rotations - High frequency, low duration motions	-Gravity Eliminated activities Supine Sidelying Prone Reclined -Sustained positions -Trunk lateral flexion positions -Low frequency movement, longer duration motions
Treatment Intervention	- 3-dimensional axial separations with rotation oscillations in extension (Figure 6)	- 3-dimensional axial separation sustained with flexion side-bend (Figure 7)
Time of day for activities	Early in am or in the afternoon after lying for 10-15 minutes	Afternoon, evening

Table 1. Application of Intervertebral Lumbar Disc Dehydration Principles.

While it is important for the clinician to understand the importance of IVD dehydration for the younger individual with discogenic low back pain, the importance of IVD rehydration cannot be overlooked. Occasionally, a person with pain from the IVD may experience pain towards the end of the day. The research involving IVD hydration is ideal because many of the recovery positions can be easily adopted in the home environment. In addition, the focus of many research articles involving the change in spine height with hydration and dehydration positions has been to identify individualized patient education suggestion. The primary limitation of research related to IVD hydration is that changes in spine height have only been demonstrated for a short duration. Ongoing research should emphasize longitudinal trials with patients applying these techniques to determine the long-term efficacy of patient education. It would also be beneficial to evaluate the continued affects of IVD hydration positions on spine height, severity of symptoms and function over an extended period of time. Moreover, much of the research has been performed on young, healthy individuals. More research involving patients with known disc degeneration will allow greater generalizability.Supine flexed postures (Figure 1) have been shown to provide a similar hydration effect on the lumbar IVD (Owens et al., 2009). Other alternative recovery positions have been shown to facilitate hydration of the lumbar IVD (Figure 2 A & B) (Gerke et al. 2010). Gerke and colleagues found that 10-15 minutes in the sidelying position will also allow a temporary amount of lumbar IVD rehydration (Gerke et al. 2010). In summary, there are a variety of positions that can be utilized to rehydrate the lumbar IVD. Utilizing a supine or sidelying position may not be available in every occupation. However, choosing the most comfortable and convenient exercise or position before an increase in pain from the lumbar IVD is felt, may prolong the amount of time an employee can tolerate work without back pain.

Fig. 1. Sustained Supine Flexed Posture.

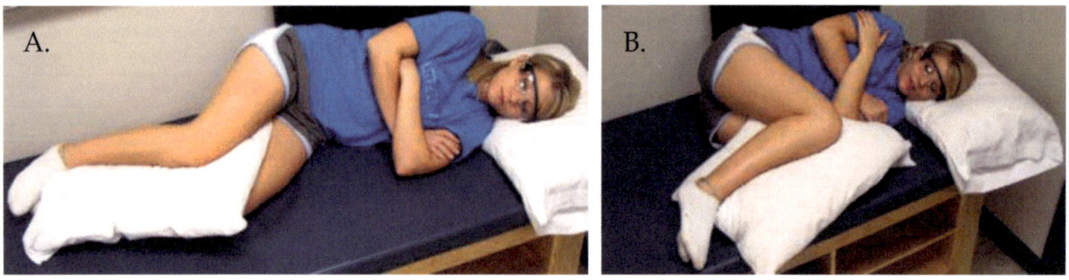

Fig. 2. A & B - Sustained Side-lying Flexed Postures.

Included in these recent findings is research performed on individuals suffering from low back pain related to nerve root compression syndrome. Simmerman and colleagues found that individuals performing aquatic traction (Figure 3) by hanging on pool noodles with 2.3 kg weight on each ankle demonstrated decreased pain as well as increased spinal height (Simmerman et al, 2011). The greatest impact that this line of research involving IVD hydration has had is that these recommendations and activities can be carried out in non clinical home based settings. Patient and individual education is the focus of this intervention strategy.

The primary limitation of research related to IVD hydration is that changes in spinal height have only been demonstrated for the short term. On- going research should emphasize longitudinal trials with subjects/patients applying these techniques to find out the efficacy of education and application over prolonged periods of time on spinal height change, symptoms and function.

Fig. 3. Aquatic Traction.

5.1 Considerations for the injured employee

Employees are often provided light duty work options. It can be common for an employee with discogenic low back pain secondary to a repetitive lifting injury or a one-time episode of excessive lifting to receive light duty restrictions which allows a more rapid return to work. In some scenarios, the employee is often counseled to perform tasks that are perceived to be easier because the task is completed in a sitting position. However, as previously mentioned, a seated position increases lumbar IVD pressure. Frequent interruptions from the seated position would allow a change in the stress distribution of body weight. Wilke et al. found that standing positions could be a better alternative than prolonged sitting (Wilke et al., 2001). While an employee may be removed from heavy lifting loads that compromised the lumbar IVD, the seated tasks could be limited in duration and recommendations for an unloading exercise or position could also used between seated tasks. Limiting the duration of sitting to 30 minutes may be beneficial to the patient experiencing pain from discogenic low back pain.

Advising employees regarding their sleeping habits can also help return the injured employee to work more comfortably. As previously discussed, the pressure in the lumbar IVD increases each hour of rest as the disc imbibes fluid. Therefore, it may be wise to prevent the lumbar IVD from absorbing water in the unloaded positions during sleep. Sleeping a shorter duration of time can be helpful to prevent the lumbar IVD from imbibing fluid and increasing pressure. In addition, activity after multiple hours of rest may also improve the exchange of waste products and nutrition. Employees on light duty may be encouraged to walk throughout the day. In most scenarios, walking can be therapeutic for an employee experiencing discogenic low back pain. Cyclic loading such as walking has been shown to diminish the effects of lumbar disc dehydration. However, according to Sizer and colleagues, pain with walking may be secondary to the attachment of the ligaments of Hoffman to the posterior fibers of the lumbar IVD (Sizer et al., 2002). These fibers can be loaded with tension from the sciatic nerve through hip flexion, knee extension and ankle dorsiflexion (Gilbert et al., 2007). These lower extremity positions can tension the sciatic nerve as often experienced while walking. Hip flexion can be diminished with shorter steps. If walking is an activity that provokes symptoms, ambulating with shorter steps should be considered when the employee notices discomfort.

In summary, to promote optimal nutrition of the lumbar intervertebral disc, repetitive activities such as walking should be encourage in the early morning to dehydrate the disc, while sitting with support or reclined or lying is more advisable by mid and end of the day to re-hydrate the disc. Frequent short breaks from static loaded working positions are advised at least every couple of hours to stimulate fluid diffusion throughout the disc (Trinkoff et al., 2006). Counseling employees with discogenic low back can be challenging. Simple recommendations such as avoiding bending or lifting early in the morning may have a significant impact on an individual's recovery.

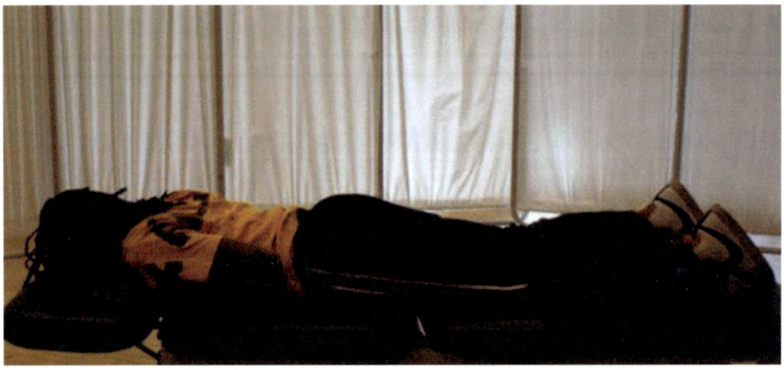

Fig. 4. Repetitive Extension in Prone.

Fig. 5. Repetitive Extension in Prone.

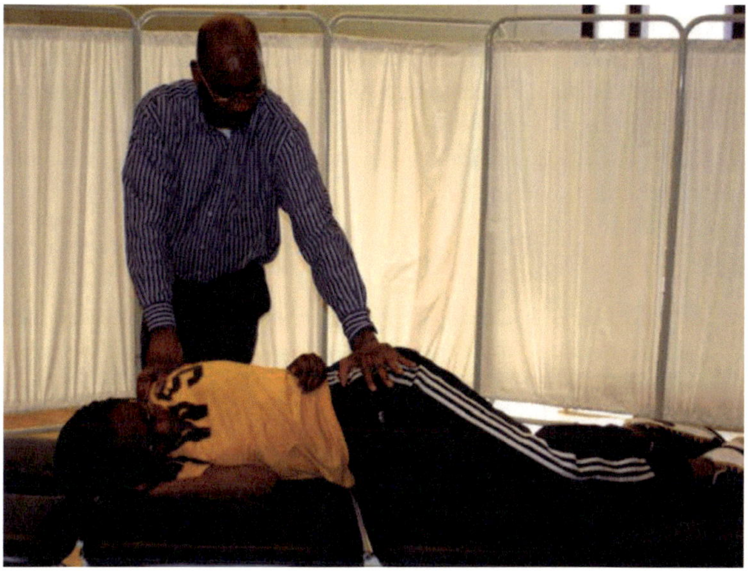

Fig. 6. 3-Dimensional Axial Separation with Rotation Oscillations in Extension.

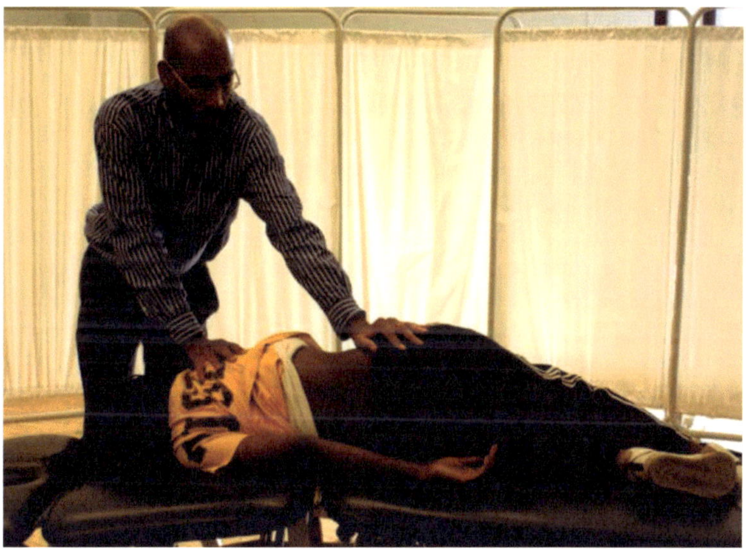

Fig. 7. 3-Dimensional Axial Separation Sustained with Flexion and Side-bending.

6. Conclusion

Low back pain has a negative financial and medical burden on society, with an estimated 80% of individuals experiencing an acute episode of low back pain at some period in their lives. Chronic low back pain is the leading cause of work-related disability for individuals under the age of forty five (Buckwalter, 1995). It is critical for healthcare professionals to have knowledge of the anatomical and biomechanical contributions to low back pain. Identifying a specific structural cause of low back pain, despite the controversy that exists, is critical for prevention and management strategies. The lumbar IVD is the primary biomechanical restraint of motion at the vertebral segment, and has been shown to play a primary role in low back pain. Intervention strategies that influence the hydration of the lumbar IVD can play beneficial role in the management of work related low back pain.. An evidenced based approach that considers these factors can be helpful in the management of work related low back pain.

7. References

Adams, M.A., McNally, D.S., & Dolan P. (1996). 'Stress' distributions inside intervertebral discs. The effects of age and degeneration. *Journal of Bone Joint Surgery*. Vol.78, No.6, pp. 965-972.

Adams, M.A., Freeman, B.J., Brian, J.C., Morrison, H.P., Nelson, I.W., & Dolan, P. (2000). Mecahnical initiation of intervertebral disc degeneration. *Spine*. Vol.25, No.13, pp. 1625-1636.

Battié, M.C., Videman, T., Kaprio, J., Gibbons, L.E., Gill, K., Manninen, H., Saarela, J., & Peltonen, L. (2009). The Twin Spine Study: contributions to a changing view of disc degeneration. *Spine Journal*. Vol.9, No.1, pp. 47-59.

Boos, N., Rieder, R., Schade, V., Spratt, K.F., Semmer, N., & Aebi, M. (1995). The diagnostic accuracy of magnetic resonance imaging, work perception, and psychosocial factors in identifying symptomatic disc herniations. *Spine*. Vol.20, pp. 2613-2625.

Boos, N., Weissbach, S., Rohrbach, H., Weiler, C., Spratt, K.F., & Nerlich, A.G. (2002). Classification of age-related changes in lumbar intervertebral discs: 2002 Volvo Award in basic science. *Spine*. Vol.27, No.23, pp. 2631-2644.

Buckwalter, J.A. (1995). Aging and degeneration of the human intervertebral disc. *Spine*. Vol.20, No.11, pp. 1307-1314.

Cinotti, G., De Santis, P., Nofroni, I., & Postacchini, F. (2002). Stenosis of lumbar intervertebral foramen: anatomic study on predisposing factors. *Spine*. Vol.27, No.3, pp. 223-229.

Coppes, M.H., Marani, E., Thomeer, R.T., & Groen, G.J. (1997) Innervation of "painful" lumbar discs. *Spine*. Vol.22, No.20, pp. 2342-2349; discussion 2349-2350.

Cook, C., Brismee, J.M., & Sizer, P.S. (2006). Subjective and objective descriptors of clinical lumbar spine instability: a Delphi study. *Manual Therapy*. Vol.11, No.1, pp. 11-21.

Crock, H.V., & Goldwasser, M. (1984). Anatomic studies of the circulation in the region of the vertebral end-plate in adult Greyhound dogs. *Spine*. Vol.9, No.7, pp. 702-706.

DePalma, M.J., Ketchum, J.M., & Saullo, T. (2011). What is the source of chronic low back pain and does age play a role? *Pain Med*. Vol.12, No.2, pp. 224-233.

Deyo, R.A., & Phillips, W.R. (1996). Low back pain. A primary care challenge. Spine. Vol.21, No.24, pp. 2826-2832.

Edwards, W.T., Ordway, N.R., Zheng, Y., McCullen, G., Han, Z., & Yuan, H.A. (2001). Peak stresses observed in the posterior lateral anulus. *Spine*. Vol.15, No.26, pp. 1753-1759.

Eklund, J.A., & Corlett, E.N. (1984). Shrinkage as a Measure of the Effect of Load on the Spine. Spine, Vol. 9, pp. 189-194.

Freemont, A.J., Peacock, T.E., Goupille, P., Hoyland, J.A., O'Brien, J., & Jayson, M.I. (1997). Nerve ingrowth into diseased intervertebral disc in chronic back pain. *Lancet*. Vol.350, No.9072, pp. 178-181.

Fryer J.C., Quon J.A., Smith F.W. (2010). Magnetic Resonance Imaging and Stadiometric Assessment of the Lumbar Discs after Sitting and Chair-Care Decompression Exercise: A Pilot Study. Spine Journal, Vol. 10, pp. 297-305.

Gerke, D.A., Brismee, J.M., Sizer, P.S., Dedrick, G.S., & James, C.R. (2011). Change in Spine Height Measurements Following Sustained Mid-Range and End-Range Flexion of the Lumbar Spine. *Applied Ergonomics*, Vol.42, pp. 331-336.

Gilbert, K.K., Brismee, J.M., & Collins DL. (2007). 2006 Young Investigator Award Winner: lumbosacral nerve root displacement and strain: part 2. A comparison of 2 straight leg raise conditions in unembalmed cadavers. *Spine*. Vol.32, No.14, pp. 1521-1525.

Helander, M.G., & Quance, L.A. (1990). Effect of Work-Rest Schedules on Spinal Shrinkage in the Sedentary Worker. *Applied Ergonomics*, Vol.21, pp. 279-284,.

Holm, S., Maroudas, A., Urban, J.P., Selstam, G., & Nachemson, A. (1981). Nutrition of the intervertebral disc: solute transport and metabolism. *Connective Tissue Research*. Vol.8, No.2, pp. 101-119.

Kanlayanaphotporn, R., Williams, M., Fulton, I., & Trott, P. (2002). Reliability of the Vertical Spinal Creep Response Measured in Sitting (Asymptomatic and Low-Back Pain Subjects). *Ergonomics*. Vol.45, pp. 240-247.

Kalichman, L., & Hunter, D.J. (2008). The genetics of intervertebral disc degeneration. Familial predisposition and heritability estimation. *Joint Bone Spine*. Vol.75, No.4, pp. 383-386.

Kourtis, D., Magnusson, M.L., Smith, F., Hadjipavlou, A., & Pope, M.H. (2004) Spine Height and Disc Height Changes As the Effect of Hyperextension Using Stadiometry and MRI. *Iowa Orthopaedic Journal*. Vol.24, pp. 65-71.

Leivseth, G., & Drerup, B. (1997). Spinal Shrinkage During Work in a Sitting Posture Compared to Work in a Standing Posture. *Clinical Biomechanics*. Vol.12, pp. 409-418.

Magnusson, M., & Pope, M.H. (1996). Body height changes with hyperextension. *Clinical Biomechanics*. Vol.11, pp. 236-238.

Magnusson, M.L., Aleksiev, A.R., Spratt, K.F., Lakes, R.S., & Pope, M.H. (1996). Hyperextension and Spine Height Changes. *Spine*. Vol. 21, pp. 2670-2675.

McMillan, D.W., Garbutt, G., & Adams, M.A. (1996). Effect of sustained loading on the water content of intervertebral discs: implications for disc metabolism. *Annals of Rheumatic Disorders*. Vol.55, No.12, pp. 880-887.

McNally, D.S., Shackleford, I.M., Goodship, A.E., & Mulholland, R.C. (1996). In vivo stress measurement can predict pain on discography. *Spine*. Vol.21, No.22, pp. 2580-2587.

Morinaga, T., Takahashi, K., & Yamagata. (1996). Sensory innervation to the anterior portion of lumbar intervertebral disc. *Spine*. Vol.21, No.16, pp. 1848-1851.

Nordin, M., & Frankel, V.H. (1989). Basic biomechanics of the musculoskeletal system. 2nd ed. ed. Philadelphia: Lea and Febiger.

Owens, S.C., Brismee, J.M., Pennell, P.N., Dedrick, G.S., Sizer, P.S., & James, C.R. (2009). Changes in Spinal Height Following Sustained Lumbar Flexion and Extension Postures: a Clinical Measure of Intervertebral Disc Hydration Using Stadiometry. *Journal Manipulative Physiologic Therapeutics*. Vol.32, pp. 358-363.

Paajanen, H., Lehto, I., Alanen, A., Erkintalo, M., & Komu, M. (1994). Diurnal Fluid Changes of Lumbar Discs Measured Indirectly by Magnetic Resonance Imaging. *Journal of Orthopaedic Research*. Vol.12, pp. 509-514.

Pennell, P.A., Owens, S.C. Brismee, J.M, Dedrick, G., James, C.R., & Sizer, P.S. (2012). Inter-tester and intra-tester reliability of a clinically based spinal height measurement protocol. Vol.1, No.2, pp. 1-4.

Rajasekaran, S., Babu, J.N., Arun, R., Armstrong, B.R., Shetty, A.P., & Murugan, S. (2004). ISSLS prize winner: A study of diffusion in human lumbar discs: a serial magnetic resonance imaging study documenting the Influence of the endplate on diffusion in normal and degenerate discs. *Spine.* Vol.29, No.23, pp. 2654-2667.

Schmidt, T.A., An, H.S., Lim, T.H., Nowicki, B.H., & Haughton, V.M. (1998). The stiffness of lumbar spinal motion segments with a high-intensity zone in the anulus fibrosus. *Spine.* Vol.23, No.20, pp. 2167-2173.

Sehgal, N., & Fortin, J.D. (2000). Internal disc disruption and low back pain. Pain Physician. Vol.3, No.2, pp. 143-157.

Selard, E., Shirazi-Adl, A., & Urban, J.P. (2003). Finite element study of nutrient diffusion in the human intervertebral disc. *Spine.* Vol.28, No.17, pp. 1945-1953; discussion 1953.

Simmerman, S.M., Sizer, P.S., Dedrick, G.S., Apte, G.G., & Brismee, J.M. (2011). Immediate changes in spinal height and pain after aquatic vertical traction in patients with persistent low back symptoms: a cross over clinical trial. *Journal of Physical Medicine and Rehabilitation.* Vol.3, No.5, pp. 447-457.

Sizer, P.S., Phelps, V., Dedrick, G., & Matthijs, O. (2002). Differential diagnosis and management of spinal nerve root-related pain. *Pain Practice.* Vol.2, No.2, pp. 98-121.

Sizer, P.S., Phelps, V., & Matthijs, O. (2001). Pain generators of the lumbar spine. *Pain Practice.* Vol.1, No.3, pp. 255-273.

Snook, S.H., Webster, B.S., McGorry, R.W., Fogleman, M.T., & McCann, K.B. (1998). The reduction of chronic nonspecific low back pain through the control of early morning lumbar flexion. *A randomized controlled trial.* Vol.23, No.23, pp. 2601-2607.

Trinkoff, A.M., Le, R., Geiger-Brown, J., Lipscomb, J., & Lang, G. (2006). Longitudinal relationship of work hours, mandatory overtime, and on-call to musculoskeletal problems in nurses. *American Journal of Industrial Medecine.* Vol.49, No.11, pp. 964-71.

Tyrrell, A.R., Reilly, T., & Troup, J.D. (1985). Circadian Variation in Stature and the Effects of Spinal Loading. *Spine.* vol.10, pp. 161-164.

Urban, J.P., & McMullin, J.F. (1988). Swelling pressure of the lumbar intervertebral discs: influence of age, spinal level, composition, and degeneration. *Spine.* Vol.13, No.2, pp. 179-187.

Urban, J.P., & Roberts, S. (2003). Degeneration of the intervertebral disc. *Arthritis Research Therapy.* Vol.3, pp. 120-130.

U.S. Department of Labor, Bureau of Labor Statistics. (2005). Nonfatal occupational injuries and illnesses requiring days away from work. Accessed September 5, 2007, at: http://www.bls.gov/news.release/pdf/osh2.pdf.

Wilke, H., Neef, P., Hinz, B., Seidel, H., & Claes, L. (2001). Intradiscal pressure together with anthropometric data--a data set for the validation of models. *Clinical Biomechanics.* Vol.16, pp. S111-126.

Zhao, F., Pollintine, P., Hole, B.D., Dolan, P., & Adams, M.A. (2005). Discogenic origins of spinal instability. *Spine.* Vol.30, No.23, pp. 2621-2630.

4

A Comparison of Software Tools for Occupational Biomechanics and Ergonomic Research

Pamela McCauley Bush*, Susan Gaines, Fatina Gammoh and Shanon Wooden
Ergonomics Laboratory,
Department of Industrial Engineering and Management Systems,
University of Central Florida, Orlando, FL
USA

1. Introduction

The purpose of this study was to evaluate and compare commercially available software tools in ergonomics and biomechanics research. The project provides a survey of select biomechanical software tools and also gives a detailed analysis of two specialized packages, 3DSSPP and JACK as well as examples of applications where one or the other may be better suited. A summarized comparison of these two packages is provided.

Three research projects in the Ergonomics Laboratory at the University of Central Florida were used to evaluate the software tools in this study. This study, entitled *A Human-Centered Assessment of Physical Tasks of First Responders in High Consequence Disasters*, looks at three of the physical tasks associated with first responders in disaster management (i.e., emergency management and response). The third case is a preliminary study to analyze the biomechanics associated with interactive gaming. The associated output not only provides a direct comparison of the two software tools, but also provides recommendations for the preferred simulation tool for appropriate biomechanical analysis for each of the three projects. The results identify the physical tasks which may place subjects at risk of physical injury and possible cumulative trauma disorders (i.e. work related musculoskeletal disorders (WMSD)). The results of the simulation analysis can be useful to researchers in assessing risks, developing worker training, selecting appropriate personal protective equipment, and recommending ergonomic interventions to mitigate risks.

For each of the three research projects being evaluated, select task elements were identified for evaluation. The task elements (or activities) selected represent a cross-section of typical physical tasks performance in physically intensive task performance (i.e. load lifting, carrying weight, or awkward posture). These tasks were simulated from photographs taken during actual task performance or still photos taken from videos. For the solid waste

* Corresponding Author

collection project, the worker was observed lifting and emptying a full canister, and moving an empty canister back into place. For the disaster management research, three tasks commonly associated with first responders were evaluated including victim extraction, supply distribution, and moving the injured. The tasks were simulated by positioning virtual models in the same postures as workers and estimating the loads. Variables which were considered included uneven ground in which workers must work, lifting loads, and body and limb postures. Variables which could not be simulated using the software tools included temperature, humidity, physical fatigue, mental stress, and chemical, biological and environmental hazards. The interactive gaming project involved observing a subject playing a controller-free video game and simulating some of the postures that were commonly performed during game play.

Software developed by the University of Michigan, 3DSSPP, was used to assess tasks from all three research projects. The 3DSSPP results were used to evaluate the loads, balance and stresses on the virtual humans. The same tasks were evaluated with the JACK software, developed by Siemens Corporation. The summary reports generated by each of the software tools were compared and analyzed for each project.

2. History and significance

The comparison of software tools for biomechanical analysis is an important aspect for understanding the most applicable tools for a given research project. In a review of the literature, few studies were identified that performed an analysis of the different features of comparable biomechanical modeling software. The growth of computer based analysis tools, dictates a need for the unbiased research community to provide analyses that can offer objective feedback on the use of these analysis tools.

Three environments that contain potentially hazardous postures in ergonomics and biomechanics were identified. These three projects are ongoing research efforts in the Ergonomics Laboratory at the University of Central Florida and provided an opportunity for a comparative study of related software products. Below are brief summaries of these projects.

2.1 Ergonomic study in solid waste collection

Municipal Solid Waste collection is a necessary activity all around the world and is associated with occupational injuries due to ergonomic risk factors including lifting, heavy load handling, awkward postures, long task durations and high levels of repetition. In the past, waste has been collected manually from customers, and has often resulted in frequent injuries to the workers. Technological development has introduced automated and semi-automated collection systems that, according to manufacturer's claims, enhance worker safety, collection productivity while at the same time reducing workers compensation claims. Thus, such advantages should balance increases in equipment cost; however, some experts suggest that for automated and semi-automated waste collection systems the capital, operating and maintenance costs are higher than costs associated with manual collection.

From the published literature, it was noticed that relatively little research has been published on ergonomics and safety in the manual waste collection industry. Additionally, the field is lacking a comprehensive study that assesses and compares the ergonomic and biomechanics issues associated with waste collection at varying levels of automation including manual, semi-automatic and automatic. This study will fill the research gap by providing an ergonomic and biomechanics assessment of the three primary approaches to waste collection.

The study utilized observational analysis, laboratory analysis and a review of historical data, where surveys were conducted for solid waste collectors and safety personnel of different waste companies in Orlando, Florida to understand the factors affecting waste collectors' safety. The focus will be on the type of waste collection tasks performed in residential communities,

Ergonomics and biomechanics evaluation techniques included postural analysis, lifting analysis, assessment of musculoskeletal risk and holistic assessment of occupational risks for workers at all three levels of automation. A detailed review of the injury data collected by the U.S. Bureau of Labor Statistics (BLS) was performed to evaluate the nature and frequency of injury incidents over time in solid waste collection field. The study will establish a foundation for additional research and recommendations for mitigating risks at all levels of task performance.

2.2 A human-centered assessment of physical tasks of first responders in high consequence disasters

This human-centered study is the initial step in developing a methodology to categorize and analyze physical tasks performed by first responders in high consequence disasters from a human factors' and biomechanics perspective. Four key phases of Disaster Management include preparedness, response, recovery, and mitigation. The tasks analyzed in this study occur in the response phase. The software tools, 3DSSPP and JACK, allowed evaluation of biomechanical risks associated with the tasks performed n the response phase. For comparative purposes the physical tasks evaluated can be partitioned into three categories: 1) Victim extraction, 2) Moving of injured, and 3) Distribution of supplies (food, water, or temporary housing supplies). Photographs of emergency workers and volunteers fulfilling these roles were retrieved from past disasters and subject matter experts. Additionally, task activities and related postures and load handling was simulated in the software environment.

The volume of rescue workers in a high consequence disaster is difficult to quantify. While professionally trained rescuers such as firemen and policemen will provide aid, according to Kano, Siegel and Bourque (2005), "It is (also) recognized that members of the lay public are often the actual 'first responders' in many disaster events." Issues related to mental stress of witnessing widespread death and devastation have been widely researched with regard to first responders. As a result of the World Trade Center Disaster in 2001, first responder health problems related to pulmonary issues due to ingested dust and particles at disaster site have also been well-documented. Personal protective equipment (PPE) is endorsed by the Occupational Safety and Health Administration (OSHA) based on data from previous disasters; however, recommended PPE equipment tends to be in response to environmental, biological or chemical risks. However specific biomechanical risks have not been widely studied among responders in high consequence emergency response.

Therefore, potentially valuable technology and personal protective equipment (PPE) such as lifting aids or back belts used in lifting tasks are missing from disaster PPE recommendations. The frequency of weather-related disasters has increased in the past ten years. From 1980 through 2009, there have been 96 weather-related disasters in which overall damages reached or exceeded $1 billion per event (NCDC). Scientists theorize the increase is related to global warming. Whatever the cause, it is clear that the frequency with which disaster workers and volunteers will need to provide aid will continue to increase. The lack of training and literature with regard to mitigation of risks to first responders as related to physical tasks, points to the need for more research in this area.

Research has focused on mental health risks such at Post-Traumatic Stress Disorder (PTSD), environmental risks such as chemicals, electrical risks due to downed power lines, and biological hazards which include "insect bites/stings, mammal/snake bites, and exposure to molds and other biological contaminants as a result of water damage, and sewage infiltration in low-lying areas". (Stull, 2006) Despite the lack of research and literature regarding risks and injuries of first responders as a result of the physical tasks performed, back injuries account for 31% of all workers' compensation claims in the United States. This fact alone indicates the need to study rescue worker safety with regard to the physical tasks performed and subsequent risks incurred by carrying out these tasks.

If physical tasks can be categorized and evaluated for risk utilizing software tools for simulation, researchers can identify those tasks which place rescue workers at greatest risk. Once these tasks are identified, collection of real-time data from disaster sites can be collected and analyzed. These results can be compared and validated by recreation of the tasks in a laboratory setting and analysis with the software simulation tools. Action in the form of enhancements to training and additional PPE recommendations can be taken to reduce the risks to these workers. Ultimately, both victims and responders will benefit from having a healthier work force that can provide faster and more efficient response, further preserving lives and expediting rescues.

2.3 Biomechanical assessment of postures associated with Interactive gaming

The advent of movement and gesture-based video gaming systems such as the Nintendo Wii, Playstation Move and Xbox Kinect have recently taken the world of gaming and computer interaction to a whole new level. Rather than controlling the game with one's digits, the player's entire body can be used to control his or her actions within the gaming interface. This sort of technology introduces a new level of activity to users who were once glued to their seats during play. Conversely, this technology has raised concerns about injuries due to the overuse of the motion-controlled mode of entertainment.

The Wii system is the first of the motion-based games on the market, as it was released in 2006. Although the technology is relatively new, there have been reports of Wii-related injuries in medical literature (Collins, 2008). Injuries that were once considered athletic-related are now occurring in individuals who play in virtual games environments. This phenomenon is more likely to occur in a sedentary population participating in the activity (Barron, 2008). One study documents an emergency surgery that was performed on a 16 year-old boy for a Lateral patella dislocation; another serious case involved a 23-year-old woman who suffered a Meniscus tear as a result of playing 10-pin bowling on a Nintendo

Wii video game. Based on interviews with orthopedists and sports medicine physicians, the majority of Kinect-related injuries are not severe. Some cases that doctors have seen range from twisted knees, sprained ankles, strains, swelling, and some repetitive stress problems (Das, 2009). Patients complaining of injuries range from young children, to teens, to young adults, and even elderly adults. A common issue with these injuries is that players do not realize that full force and motion is not required for the game to acknowledge the action. Instead of a minuscule jump, the user might perform a full leap. If a swing action is required, many users may force a full swing when the game may only require a flick of the wrist. Some medical professionals suggest that sports injuries and cumulative trauma disorders may be likely directions for the types of injuries that may occur due to interactive game use (Barron, 2008).

Presently, interactive gaming technology is new and little published scientific research exists, particularly in the area of biomechanics. However, this poses an excellent opportunity to identify the possible risks associated with the use and over use of the systems. Performing ergonomic and biomechanical evaluations of these motion-activated games could benefit customers, manufacturers, and medical professionals, alike.

The objective of this study is to employ human modeling and simulation tools to identify potential hazards associated with some of the awkward postures exhibited during game play. The 3DSSPP and JACK programs are mainly used in analyzing occupational manual material-handling tasks. In this study these software tools will be used to simulate postures of the subject to determine if these products can go beyond occupational applications to support healthy biomechanics in design of a recreational product such as an interactive game.

3. Literature review

An internet search of software available for biomechanical analysis resulted in a significant number of options. The majority of the software offered online tends toward biomechanical evaluation for sports applications. Several of the software packages claim to accept user-provided video for analysis, but demonstrations of the software have established that these tools cannot readily accept typical user video. The videos to be used as input for most of the software must be made at a particular resolution, recorded from certain angles or be in a format which requires special, sometimes expensive, hardware. The general survey of biomechanical analysis software reveled that there are three main categories of software: 2D Video Analysis, 3D Motion Capture Analysis, and Human Modeling and Simulation programs. Several of these packages are discussed below.

3.1 2D video analysis

3.1.1 MotionView

MotionView video analysis software for sports is video coaching software that advertises it can accept input from any video camera and computer to analyze or coach sports and motion; however, the makers require the user to purchase special equipment from them to capture the videos. This software is used primarily for sports evaluation. "MotionView video analysis software for sports delivers features typically found in video analysis and swing analysis software costing much more. MotionView video analysis software for sports is golf swing analysis software, bowling video analysis software, and tennis stroke video analysis software! Improve any athletic skill with our video analysis software." The MotionSuite complete package costs $1180 http://www.allsportsystems.com/

3.1.2 ProAnalyst professional

ProAnalyst software initially seemed to be a promising tool in which user-supplied video could be downloaded and analyzed. While it does accept some user videos, there are restrictions with regard to the quality of resolution and the camera angles from which the video can be taken (side views only). Again, ProAnalyst® is used primarily for evaluation of sports; however, there are applications in the aerospace industry, such as tracking missile paths and speeds. ProAnalyst advertises that it "is the world's premier software package for automatically measuring moving objects with video. ProAnalyst allows you to import virtually any video and quickly extract and quantify motion within that video. Used extensively by NASA, engineers, broadcasters, researchers and athletes, ProAnalyst is the ideal companion software to any consumer, scientific and industrial video camera, and vice versa. With ProAnalyst, any video camera becomes a non-contact test instrument. ProAnalyst allows users to measure and track velocity, position, size, acceleration, location and other characteristics." ProAnalyst does provide the ability to export data into graphical formats, but it did not prove to be as user-friendly for occupational evaluation, and it required cameras which recorded at a higher resolution than the typical home video camera. The ProAnalyst Professional Edition, Ultimate Bundle costs $9595. http://www.xcitex.com/html/proanalyst_applications_examples.php

3.1.3 MaxTRAQ 2D

MaxTRAQ 2D can use a standard camcorder to high speed camera for input. This program also features a manual or automatic digitizer that can be used to extract kinematic properties from standard AVI files. This feature is useful when markers cannot be placed on the subject. MaxTRAQ includes tools to measure distances, angles, center of mass, etc. The price of this software is $695. http://www.innovision-systems.com

3.2 3D motion capture

3.2.1 Visual3D professional

C-Motion-Visual3D biomechanical analysis software is marketed as being "used for performance analysis and movement assessments." The applications appear to be more pertinent to the medical community. This software does require an existing motion capture system. For this system, cameras are not directly supported. Video data must be preprocessed into a digital format that Visual3D can process and analyze. Depending on the Motion capture setup, this may require additional software. This system also reports that data from Force Platform and EMG analog devices can be synchronized with the video. The cost for the Visual3D Professional (with Real-Time Biofeedback, Relational Database Export, Inverse Kinematics, 4-user License) is $15995. http://www.c-motion.com/products/visual3d.php

3.2.2 MaxPRO

MaxPRO is a motion caption and analysis product that can be used for research, clinical/Physical Therapy, biomechanics, sports, ergonomics, industrial/automotive, lab course/teaching. MaxPro offer 3D motion analysis without the use of a proprietary camera. This system can utilize standard camcorders to high speed, high resolution cameras. Some of the features of this program include up to a 32-camera configuration, tracking for up to 255 markers, video overlay, and graphs. The tools available can detect angles, velocity, and acceleration. This software price is listed as $4,995. http://www.innovision-systems.com

3.2.3 SIMM

SIMM Biomechanics Software Suite by MusculoGraphics "enables a detailed analysis, documentation and comparison of posture and movements"; however, it requires specialized software for simulation. It does not utilize video download features. http://www.musculographics.com/

3.2.4 ProAnalyst 3-D professional

ProAnalyst 3-D Professional Edition uses video from two cameras to create a 3D analysis tool. The system requires a special calibration tool that allows for the user to "drag and drop two calibration images in the 3-D Manager window and let ProAnalyst automatically determine the positions of the cameras. Then, add analyzed videos and allow ProAnalyst to calculate where your tracked objects are in 3-dimensional space. Finally, export your data to a fully customizable 3-axis plot and save a new video showing your analyzed event from any angle." The cost of this package is $14995.

http://www.xcitex.com/html/proanalyst_applications_examples.php

3.3 Modeling and simulation

3.3.1 3DSSPP

The 3DSSPP (3D Static Strength Prediction Program) was developed by The Center for Ergonomics at the University of Michigan College of Engineering. This program can be used in analyzing manual materials - handling tasks. Ergonomists, engineers, therapists and researchers, may use the software to evaluate and design jobs. This program allows for users to input anthropometric data, and obtain the forces and moments computed by the program, rather than by manual calculation. In addition, the program also combines the National Institute of Occupational Safety (NIOSH) lifting data and other additional reports to identify risks associated with a particular task. This software license costs $1495 (University of Michigan).

3.3.2 JACK

JACK is a human simulation tool for populating designs with virtual people and performing human factors and ergonomic analysis. JACK is a human modeling and simulation tool. JACK, and its optional toolkits, provides human-centered design tools for performing ergonomic analysis of virtual products and virtual work environments.

JACK enables you to size human models to match worker populations, as well as test designs for multiple factors, including injury risk, user comfort, reachability, line of sight, energy expenditure, fatigue limits and other important human parameters. This software license costs $2400 www.siemens.com/tecnomatix

Fig. 1. Simulation from JACK software, Technomatix.

3.3.3 Ergowatch

Ergowatch is another computerized ergonomics package system. It consists of different ergonomic measurement tools that can help employers, ergonomists, and workers to estimate and interpret the physical loading associated with various jobs. The Ergowatch package provides the below tools for work evaluation:

1. The 4D Watbak Tool which is easy to use biomechanical modeling software, to calculate instantaneous and accumulated loads for the lower back and other major body joints, during various activities and to predict the relative risk of lower back injury
2. The NIOSH Tool which provides load limits for lifting and lowering activities (based on the 1981 and 1991 NIOSH Lifting Equations)
3. The Snook Tool: Provides load limits for lifting, lowering, pushing, pulling and carrying activities (based on the 1991 Revised Snook Tables)
4. The Physical Demands Description (PDD) Checklist Tool: Structures the description of physical movements and environmental conditions associated with a task group or job (adapted from the Ontario Ministry of Labor Physical Demands Analysis form). The cost of this package is $1500. http://www.escs.uwaterloo.ca/brochure.pdf

3.3.4 AnyBody modeling system

The AnyBody Modeling System™ is a software system for simulating the mechanics of the live human body working in a particular environment. AnyBody has applications in the auto industry, medicine, the aerospace industry, sports analysis, research, and even defense. The software runs a simulation and calculates the associated mechanical properties including individual muscle forces, joint forces and moments, metabolism, elastic energy in tendons, antagonistic muscle actions and much more. AnyBody can also import data from Motion Capture systems. The pricing was not available without a full demonstration. The company is headquartered in Sweden. http://www.anybodytech.com/index.php?id=26

A priority of this software research was to find tools which did not require a significant financial commitment, particularly with regard to specialized hardware, as that technology can be costly and often times the technology evolves quickly, rendering older generations of hardware obsolete. Due to the cost of software and required peripherals for motion analysis options, the next-best alternative for analyzing postures and loads is a simulation tool. In particular, the software utilized in this study, JACK and 3DSSPP, allowed user-supplied input for simulation of postures involved in specific tasks. A summary of the features and costs of the aforementioned software is provided in Table 1.

Software Features and Costs	MotionView	ProAnalyst Professional	MaxTRAQ 2D	Visual3D Professional	MaxPRO	SIMM	ProAnalyst 3-D Professional	3DSSPP	Jack	Ergowatch	AnyBody
2D Analysis	X	X	X	X	X	X	X	X	X	X	X
3D Analysis				X	X	X	X	X	X		X
Camera Required	X	X	X		X	X	X		OPT		OPT
Allows Import of video files			X		X						
Multiple Cameras				X	X	X	X		OPT		OPT
High Speed or High Resolution Cameras		X		X		X	X		OPT		OPT
Calibration Equipment				X	X	X	X		OPT		OPT
Limited to	X										
Existing MoCap Required				X		X			OPT		OPT
Muscle Data						X					X
System Cost	$1180	$9595	$695	$15995	$4995	N/A	$14995	$1495	$2400	$1500	N/A

Table 1. Summary of Commonly Available Biomechanical Software.

4. Methodology

The methodology used in this study can be divided into three parts: Selection of Tasks, Tools Used, and Procedures and Analysis. All three of the projects utilized video or photos of the tasks as the basis for the modeling. Some of these photos were taken by the researchers and some were retrieved from the internet. The primary methods of research included internet searches, references to Ergonomics and Biomechanics texts and course notes, search of library archives for relevant research, and creation of simulations using the software in the Ergonomics Laboratory of the University of Central Florida

The Ergonomic Study in Waste Collection also utilized observational analysis, laboratory analysis and a review of historical data, where surveys were conducted for solid waste collectors and safety personnel of different waste companies in Orlando, Florida to understand the factors affecting waste collectors' safety.

4.1 Selection of subjects

Subjects for this study were selected from a population of university students and practitioners. The subjects were selected as components of the three research projects. The tasks identified for evaluation were necessary elements of task performance for the projects and also provided an opportunity for the comparative analysis. The identified tasks contained "task elements" that were used to create simulations and generate data with regard to loads, balance, strength exertion, posture and other task performance descriptors. The tasks which were simulated are described below.

4.1.1 Solid waste collection project

4.1.1.1 Task 1: Lifting of a full waste container

Manual lifting of waste containers expose the waste collectors to severe ergonomic risks, repeating this heavy lifting several times during the day lead to musculoskeletal disorders and injuries. This task was broken down into three poses and will be explained in the analysis section.

Fig. 2. Lifting the Waste Container Task.

4.1.1.2 Task 2: Dumping of a full waste container

As per the survey that was conducted with the waste collectors, the estimated average container weight is 40 to 60 pounds. Dumping the container that is filled with waste requires awkward postures especially on the lower back region.

Fig. 3. Dumping the Garbage Container into the Back of the Truck.

4.1.2 Disaster management project

4.1.2.1 Task 1: Supply distribution

This task shows the awkward shoulder and arm angles at which supplies are sometimes lifted and moved. This is a typical first responder task.

Fig. 4. New Jersey National Guard's Response to Hurricane Katrina, Photo courtesy of pdcbank.state.nj.us

4.1.2.2 Task 2: Victim extraction

Often victim of disasters become trapped in the rubble. Rescues often require awkward and sometimes dangerous postures to keep the victim from incurring additional injury.

Fig. 5. LA Search and Rescue pull woman out of rubble 12 Jan 2010 Haiti Quake, Photo courtesy of edwardrees.wordpress.com

4.1.2.3 Task 3: Moving the injured

Keeping a victim's head and neck stationary sometimes requires an awkward position by the rescuer. In this case, other rescuers should be taking some of the load at the feet and mid-body so the rescuer does not have to support the entire weight of the victim while keeping the neck stationary.

Fig. 6. Rescuers carry injured quake victim from collapsed building, Beichuan County, China. May, 2008, Photo courtesy of nytimes.com

4.1.3 Interactive gaming project

For the observational analysis, the experimenter observed and video-recorded the subject performing the Kinect ™ Sports "Super Saver" Soccer Game.

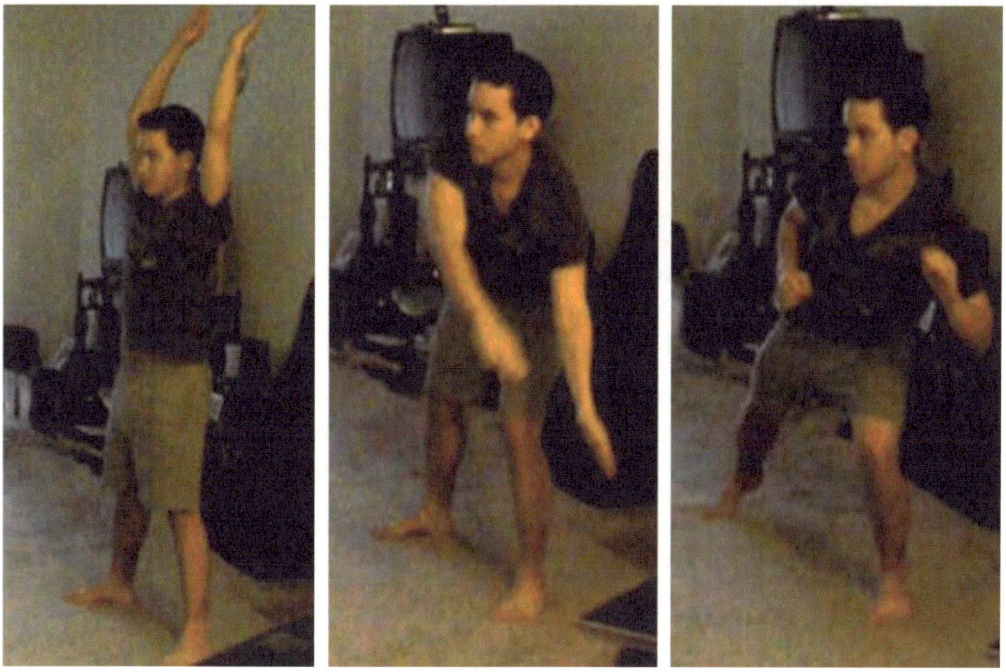

Fig. 7. Still shots of video for Tasks 1-3, respectively.

4.1.3.1 Task 1: Overhead catch

The subject reaches above his head to "catch" the soccer ball and prevent the opponent from scoring.

4.1.3.2 Task 2: Low-ball upper-limb save

The subject reaches across his body to "block" the goal. This movement involves reaching across the midline of the body, resulting in flexion, lateral bending, and rotation of the torso.

4.1.3.3 Task 3: Low-ball lower-limb save

The final pose selected involves the subject's attempt to block the ball with his foot. He extends his right leg, while putting the majority of his weight on the left side of the body.

4.2 Tools used

This study primarily required use of two software tools with which to perform biomechanical analysis. Learning the basic functionality of the JACK software required a steep learning curve. Even with extensive man-hours using the manual and the tutorials, the researchers recognize that the software was not utilized to its' maximal functionality. Vendor instructional courses would greatly enhance the users' understanding of all of the features. Despite a rudimentary use of the simulation features, usable data was generated. This data was used to analyze risks associated with the tasks included in the study. This data was also compared with the output generated for the same tasks from another simulation tool, 3DSSPP.

The following is a summary of the equipment utilized and its purpose:

4.2.1 Goniometer

In ergonomics, a goniometer is used to measure, in degrees, active or passive range of motion of applicable joints. This is pertinent to workplace design and functional reach. It can also measure progress in return of range of motion after an injury. For this study, the goniometer was used to measure the angles of limbs for the subject when recreating the solid waste collection tasks in the laboratory.

4.2.2 3D SSPP biomechanical software from the University of Michigan

3DSSPP predicts static strength requirements for tasks. The program allows user input to simulate the subject postures and loads, and use custom anthropometrics or draw from the installed tables. Output from the software includes spinal compression forces, the percentiles of humans who could perform the task, and data comparisons to NIOSH guidelines, which generate color-coded warnings. The analysis is augmented by graphic illustrations of the positions being studied (University of Michigan, 2010).

The primary feature of interest in the 3DSSPP software, for the purposes of this study, were the low back compression forces, particularly on L5/S1, the region of the spine most prone to lower back injury. These results are displayed in graphical the Summary Analysis Reports where the mark indicates if the force is acceptable (green), caution (yellow) or hazardous (red). The balance reports, moments, and strength analysis reports were also utilized.

4.2.3 Siemens PLM JACK and the task simulation builder

"JACK is a human modeling and simulation tool" (JACK) which allows user input simulate a task or environment. "Manufacturing companies in a variety of industries are addressing the human element as a key component of the design, assembly and maintenance of products". JACK utilizes a Task Simulation Builder to enable use of pre-programmed commands to instruct a human model in a virtual 3D environment. This software has a large learning curve, but once the scene is created, the computer will predict the worker movement, utilizing a library of common human movements. The human posturing features clearly incorporate research on prediction of human postures based on any change to the virtual human's posture with regard to variables including hand force exertions, foot positions, center of gravity, head position, and obstacles.

4.3 Procedure and analysis

Research including interviews, observations, and literature searches related to each of the three projects yielded preliminary elements to be considered.

For the solid waste collection project, the data collection involved interviewing employees and videotaping of a variety of tasks. The video data was uploaded into a desktop computer and viewed through Windows Movie Maker software. This program allows user to preview the video and under the Tools Tab "Take a Picture from the Preview." The video was viewed at normal speed. The user was then able to go through the video, frame by frame, and capture the exact moment to be evaluated and save it as a still photo. This allowed the user to create virtual humans and duplicate the postures and loads of the tasks.

For the disaster management project, still photos of rescue workers were downloaded from a variety of international disasters and used for analysis. The researcher was able to use the photos to create virtual humans and duplicate postures and estimate loads.

The interactive gaming study consists of an observational analysis. The experimenter interviewed the subject to determine his experience with the Xbox Kinect and other video games. Afterwards the subject's anthropometric data was collected. His height was measured with a measuring tape and his weight was measured on a digital scale. For the observational analysis, the experimenter observed and video-recorded the subject performing the Kinect ™ Sports "Super Saver" Soccer Game. Next, the video was imported into Windows Live Movie Maker to obtain freeze frames of awkward postures. Next, these snapshots were imported into Adobe Photoshop, where joint angles were determined with the measuring tool.

Once the tasks were identified and the poses selected for simulation, JACK software was used to create a virtual environment to recreate the task. The anthropometric data differed depending upon which of the virtual human models were selected. Hand loads were measured for the waste collection project and those actual loads were used for the objects in the virtual environments. For the disaster management project, the loads were estimated based on user experience. The subject in the interactive gaming project did not have a y hand load. For the purposes of this research, the human posturing techniques were the most useful. This feature allowed users to quickly posture the human model while making predictions of the next movements, based on research of actual human movements and mechanics. The postures from the photos were recreated. An example of the closely simulated posture is shown in Figures 8 through 10.

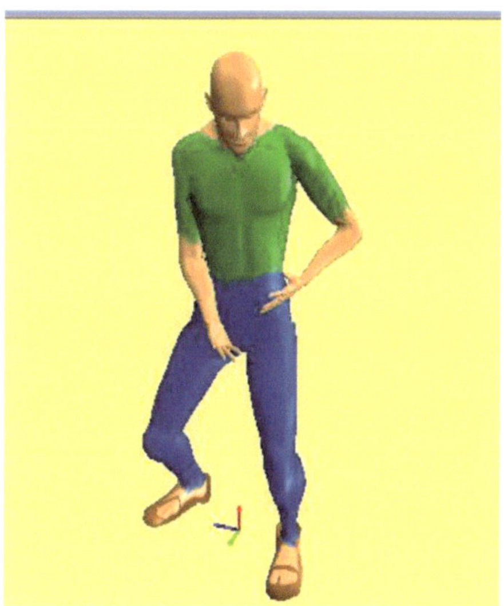

Fig. 8. 3DSSPP Simulation.

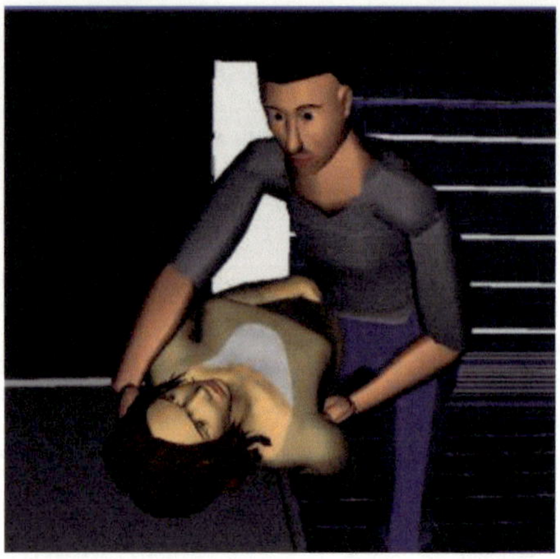

Fig. 9. JACK Simulation of Moving the injured.

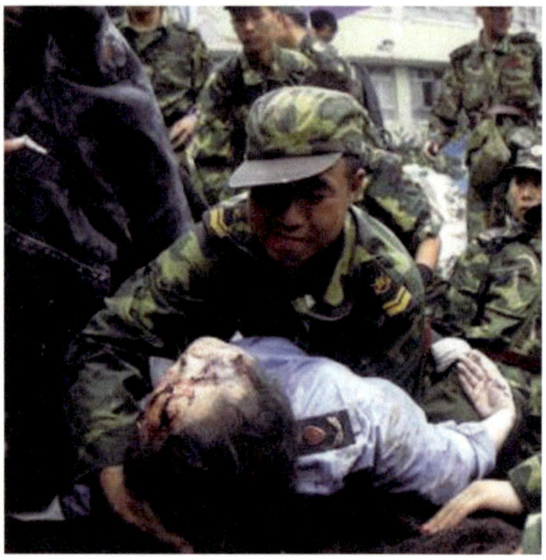

Fig. 10. Moving the Injured.

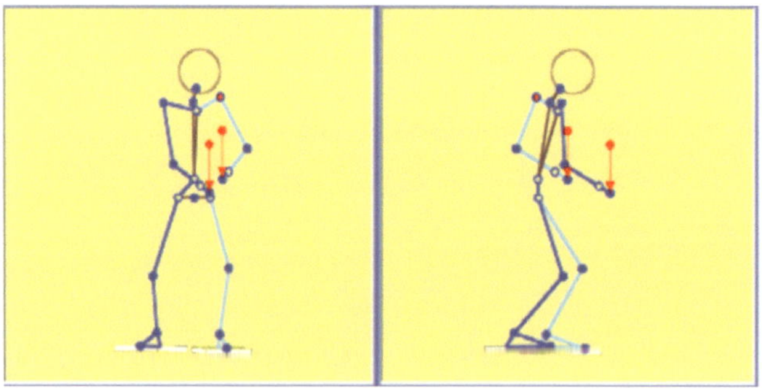

Fig. 11. 3DSSPP Simulation of Moving the Injured.

In the JACK Task Simulation Builder, once the virtual human was manipulated into the correct position, the pose was saved and used in the simulation to be sure the model retained the same pose. The software can help predict the movements either just before or after the pose or poses which are simulated. The reports of interest for this study that were generated from JACK included the joint report, the forces report, and the strength analysis report. An attempt was made to utilize the joint angle report to use the JACK values as the starting values to be used in the 3DSSPP simulation. Unfortunately, the joint angles generated by JACK are not the same angles that 3DSSPP requires for input.

The University of Michigan 3DSSPP Biomechanical software was used to analyze the poses, as well. Since 3DSSPP calculates the angles of input from the horizontal, some of the limb orientations and postures of the virtual figure had to be manually manipulated to attain a similar pose. The weights of the objects were the same as those entered in the JACK software. Hand postures were closely matched, as well. An example of the Moving the Injured task simulation in 3DSSPP is seen in Figure 11. The anthropometrics, height and weight, of the virtual figure in JACK were entered into 3DSSPP to keep the variables between the two software packages the same.

The task analysis reports in 3DSSPP predict the percentage of the population who could perform the tasks and were compared with the same percentages generated in JACK. The forces on L4/L5 were also compared. An attempt was made to compare the moments and joint angles, but the degrees of freedom allowed in manually manipulating the virtual figure in 3DSSPP made those factors inconclusive.

3DSSPP gives the Strength Limits for percent capable (percent of the population with sufficient strength) in a graphical format. The green zone is if over 99% of the population can perform the task. The yellow zone is for 25% to 99% of the population and the red zone is if less than 25% of the population can perform the tasks. JACK gives a red indicator if less than 99% of the population can perform the task. The percentages were translated from JACK and color-coded to be consistent with the red, yellow, green coding of 3DSSPP to visually clarify the results depicted in the Comparison of JACK and 3DSSPP Output Section of this paper.

5. Data analysis and results

Both 3DSSPP and JACK utilize the National Institute of Occupational Safety and Health (NIOSH) lifting guidelines to determine if loads are acceptable. With regard to evaluating whether a simulated posture falls within 'acceptable' limits, the JACK user manual states, "(JACK) Evaluates jobs in real-time, flagging postures where the requirements of a task exceed NIOSH or user-specified strength capability limits." The 3DSSPP User Manual states the following with regard to NIOSH guidelines, "NIOSH recommended limits for percent capable (percent of the population with sufficient strength) are used in the program by default. These values are documented in the Work Practices Guide for Manual Lifting (NIOSH, 1981)" (3DSSPP Manual, p. 3).

Two metrics were used as the primary tools to compare the software: The forces on L4/L5 and the strength capability of the population. The documentation and user guides of the software describe the science behind the calculation of these figures. An attempt was made to compare the moments on L4/L5. Significant variability resulted in the comparison of all of the metrics. Based on the vendor-supplied literature to the software packages, the researchers theorized that the differences in the moment and other calculations may have been due to variations in anthropometric data sources, joint angle input and exact posture replication. These topics are discussed in greater detail in the conclusion.

5.1 Static Strength Prediction percentage capable

The following quote was taken directly from the JACK user manual: "The Static Strength Prediction (SSP) tool is based on strength studies performed over the past 25 years at the University of Michigan Center for Ergonomics and augmented with data from 250 strength-related papers. A collection of strength studies is described in Occupational Biomechanics, 2nd Edition, Chaffin and Anderson, 1991…SSPP was updated for JACK v7.0 to include Wrist Strength using strength equations developed at the University of Michigan Center for Ergonomics. These equations are the same as used in the University's 3DSSPP program and were developed from an analysis of wrist and hand strength data reported in the academic literature (JACK TRAINING MANUAL, p. 18)."

While JACK bases its static strength prediction percentages from data collected at the University of Michigan, 3DSSPP was actually developed by the University of Michigan and utilizes the same population data to calculate the percentages capable for strength. The 3DSSPP Static Strength Prediction Program Version 6.0.3 User Manual (2010) states that, "Population mean strengths...are computed from empirical mean strength equations. The evaluations are based on experimental strength studies by Stobbe (1980); Shanned (1972); Burgraaff (1972); Clarke (1966); Smith and Mayer (1985); Mayer et al (1985); Kishino et al. (1985); Kumr, Chaffin, and Redfern (1985); and many others (3DSSPP Manual, p.84)."

5.2 Low Back Analysis (forces and moments on L4/L5)

The JACK User Manual discusses the Low Back Compression Analysis Tool, "the module (that) computes the spinal forces at L4/L5 utilizing the distributed moment histogram (DMH) technique for torso muscle recruitment. (JACK User Manual Version 7.0)…The Low Back Compression Analysis Tool helps evaluate the spinal forces acting on a virtual human's back. The tool tells you compression and shear forces at the L4/L5 vertebral disc, and how the compression forces compare to NIOSH recommended and permissible force limits. The results of a low back Compression analysis can be used to design or modify manual tasks to minimize the risk of low back injuries and conform to NIOSH guidelines. The tool can also pinpoint the exact moments of a lift when the compression forces on a worker's L4/L5 vertebral disc exceed NIOSH force limits. (JACK Low back Analysis Compression Tool Background C:\Program Files\Siemens\Jack_7.0\library\help\TAT_Low_back.htm)

The SSPP Manual states that, "the predicted disc compression force shown in the analysis summary can be compared to the NIOSH BCDL of 3400 Newton (3DSSPP Manual, p. 82)." This study focused on L4/L5 compression and moments. This metric is validated by the 3DSSPP manual which states, "Torso muscle moment arms and muscle orientation data for the L4/L5 level have been studied more extensively than at any other lumbar level." (3DSSPP User Manual, p. 4)

5.3 Comparison of JACK and 3DSSPP output

5.3.1 Subject 1: Solid waste collection project

5.3.1.1 Subject 1, Task 1, Pose 1: Lifting of a full waste container

Source Photo

Fig. 12. Lifting Full Waste Container.

JACK

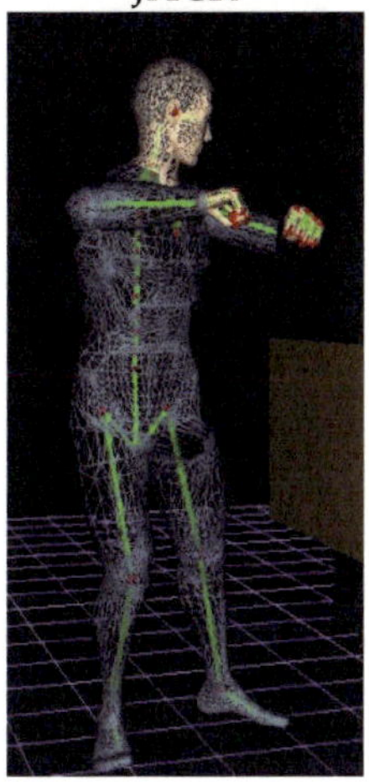

Fig. 13. JACK Simulation - Task 1, Pose 1.

3DSSPP

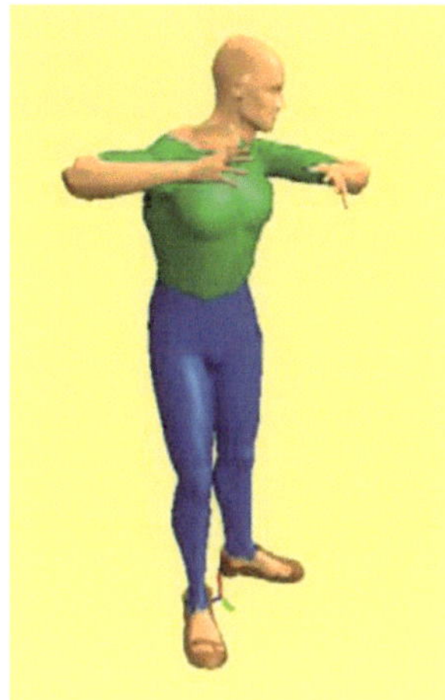

Fig. 14. 3DSSPP Simulation - Task 1, Pose 1.

	JACK	3DSSPP
L4/L5 Compression Force (N)	3203	2645

Table 2. Comparison of JACK and 3DSSPP L4/L5 Forces for Subject 1, Task 1, Pose 1 *(See report results in Figures 15 and 18).*

	JACK		3DSSPP
Joint	(% Capable)		(% Capable)
Wrist	86	41	
Elbow	100		99
Shoulder	15		92
Torso	95		91
Hip	97	77	
Knee	99		86
Ankle	78	57	

Table 3. Comparison of JACK and 3DSSPP Strength Capability Summary for Subject 1, Task 1, Pose 1 *(See report results in Figures 15 and 17).*

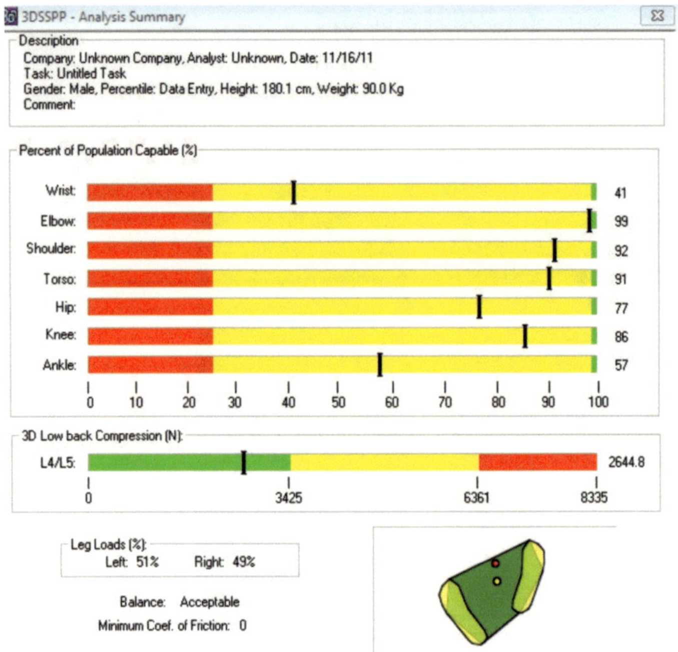

Fig. 15. 3DSSPP Summary Output, Task 1, Pose 1.

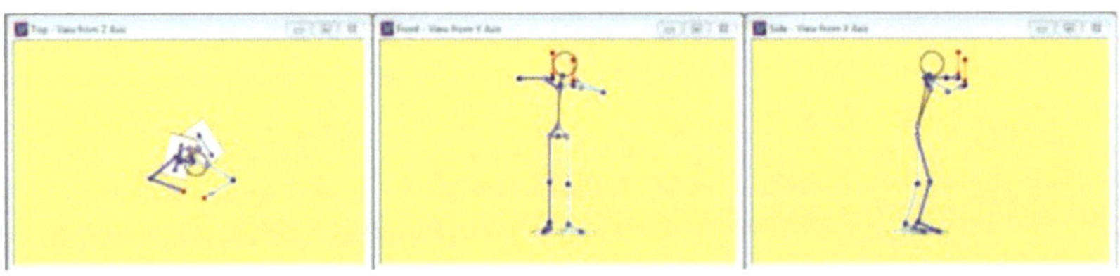

Fig. 16. 3DSSPP Simulation of Limb Angles, Task 1, Pose 1.

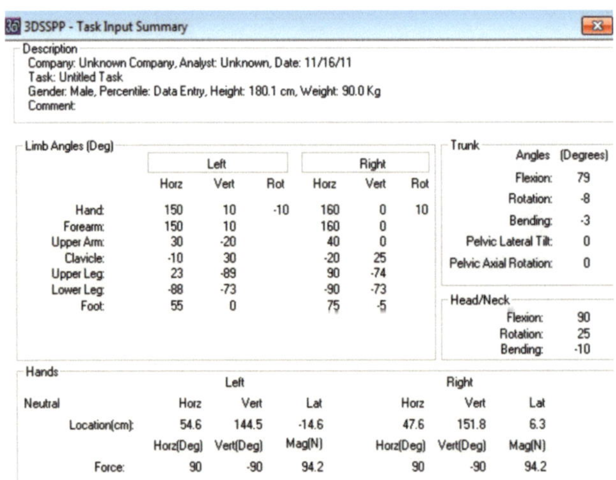
3DSSPP - Task Input Summary

Description
Company: Unknown Company, Analyst: Unknown, Date: 11/16/11
Task: Untitled Task
Gender: Male, Percentile: Data Entry, Height: 180.1 cm, Weight: 90.0 Kg
Comment:

Limb Angles (Deg)

	Left Horz	Left Vert	Left Rot	Right Horz	Right Vert	Right Rot
Hand:	150	10	-10	160	0	10
Forearm:	150	10		160	0	
Upper Arm:	30	-20		40	0	
Clavicle:	-10	30		-20	25	
Upper Leg:	23	-89		90	-74	
Lower Leg:	-88	-73		-90	-73	
Foot:	55	0		75	-5	

Trunk

Angles	(Degrees)
Flexion:	79
Rotation:	-8
Bending:	-3
Pelvic Lateral Tilt:	0
Pelvic Axial Rotation:	0

Head/Neck

Flexion:	90
Rotation:	25
Bending:	-10

Hands

Neutral	Left Horz	Left Vert	Left Lat	Right Horz	Right Vert	Right Lat
Location(cm):	54.6	144.5	-14.6	47.6	151.8	6.3
	Horz(Deg)	Vert(Deg)	Mag(N)	Horz(Deg)	Vert(Deg)	Mag(N)
Force:	90	-90	94.2	90	-90	94.2

Table 4. 3DSSPP Limb Angle Input, Task 1, Pose 1.

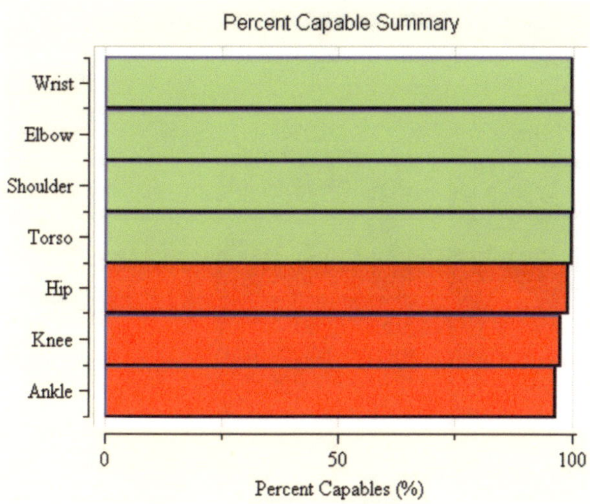

Fig. 17. JACK Output of Percent Capable, Task 1, Pose 1.

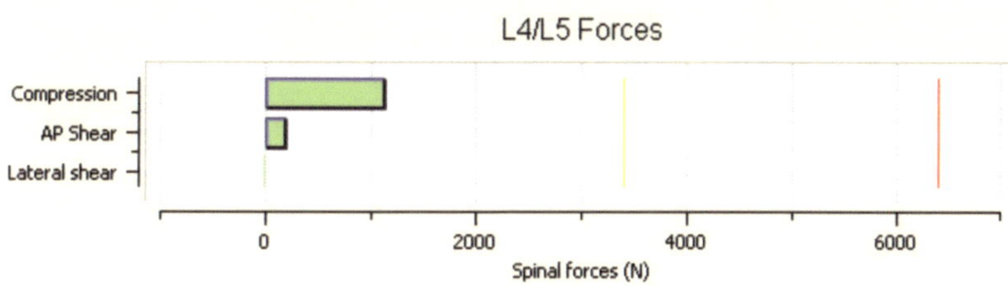

Fig. 18. JACK Output - Forces on L4/L5 for Task 1, Pose 1.

5.3.1.1.1 Analysis of Subject 1 - Task 1, Pose 1

Both packages predicted that this pose does not represent severe risk of low back injury since the compression force L4/L5 is below the NIOSH Back Compression Action Limit of 3400 N. In 3DSSPP, the force was 2645 N, while in JACK it was higher by 21%. The waste collector was not bending his torso, so it didn't require high flexion to lift the waste container; the weight of the waste container that was used in the simulation was around 19 kg. The compression force will be higher if the waste collector lifted a heavier container and bent his torso. For the percent capable, it was noted that there is a significant difference for the shoulder joint population strength between both packages, JACK indicated that only 15 % of the population will be able to perform this pose. On the other hand, 3DSSPP predicted that 92% of the population is able to perform this pose. This difference may be attributed to the manual manipulation of the postures in 3DSSPP, since this software does not provide the same flexibility to move and twist the shoulder joints as in JACK. Accordingly, for this task, JACK is more applicable to use than 3DSSPP, as it provides more flexibility to manipulate the body joints.

Analysis Recommendations:

The low back compression force of 1123.00 is below the NIOSH Back Compression Action Limit of 3400 N, representing a nominal risk of low back injury for most healthy workers.

5.3.1.2 Subject 1, Task 1, Pose 2: Lifting of a full waste container

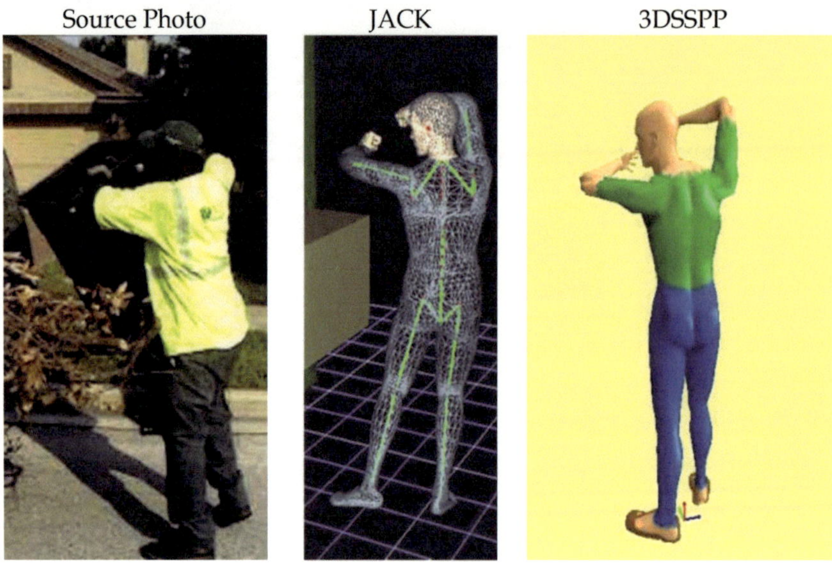

Fig. 19. Lifting Full Waste; JACK; 3DSSPP.

	JACK	3DSSPP
L4/L5 Compression Force (N)	3243	2566

Table 5. Comparison of JACK and 3DSSPP L4/L5 Forces for Subject 1, Task 1, Pose 2 *(See report results in Figures 20 and 23).*

	JACK	3DSSPP
Joint	(% Capable)	(% Capable)
Wrist	92 63	
Elbow	44	68
Shoulder	2	89
Torso	99	90
Hip	97 72	
Knee	99	75
Ankle	80 59	

Table 6. Comparison of JACK and 3DSSPP Strength Capability Summary for Subject 1, Task 1, Pose 2 *(See report results in Figures 20 and 22).*

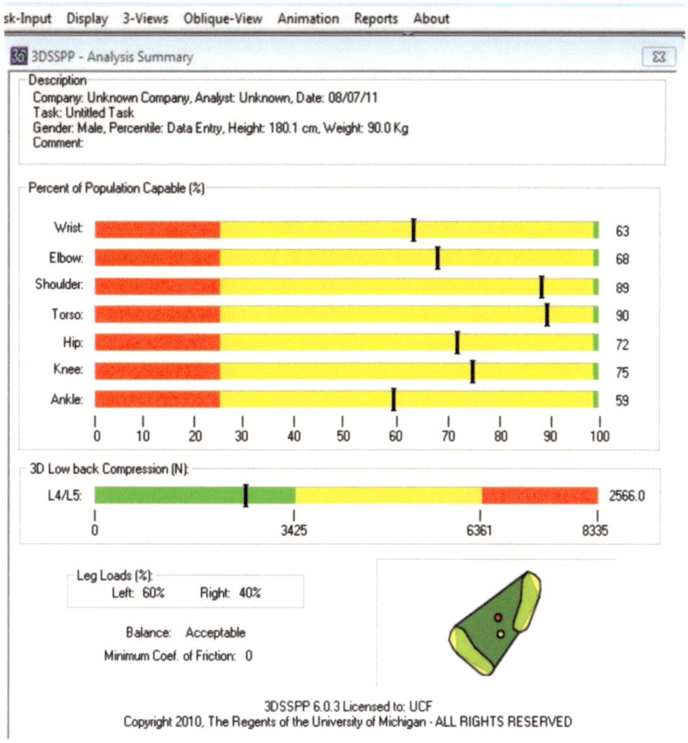

Fig. 20. 3DSSPP Summary Output, Subject 1, Task 1, Pose 2.

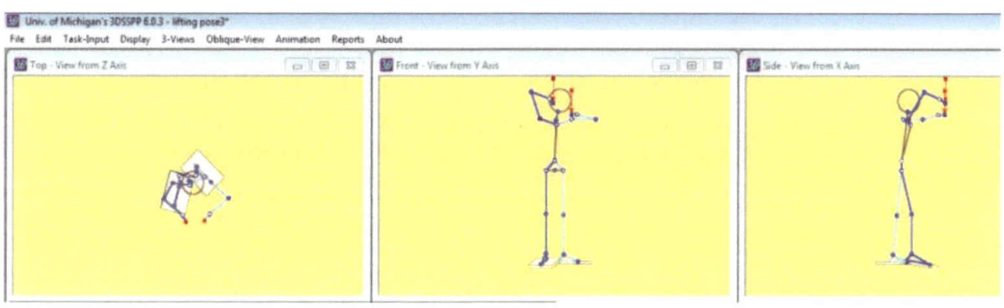

Fig. 21. 3DSSPP Simulation of Limb Angles, Subject 1, Task 1, Pose 2.

5.3.1.2.1 Analysis of Subject 1 - Task 1, Pose 2

Although 3DSSPP and JACK indicated that the compression force L4/L5 for this pose acceptable; the force is high and close to the Back Compression Action Limit. In 3DSSPP the force was 2566 N while in JACK it was higher by 26%. For the percent capable, similar to the previous pose, it was noticed that there is a significant difference between both packages, for the population strength in the shoulder joint; JACK indicated that only 2 % of the population will be able to perform this pose, while 3DSSPP indicated that 89% of population will perform this pose. According to the observational analysis and the videos, it was noticed that this task requires lifting the garbage container by elevating the shoulder and upper arms at high distance, representing an awkward posture. It was easier to manipulate and rotate the shoulder and the upper arm joints on JACK than 3DSSPP. For the other joints, both packages indicated that they would fall within the yellow zone.

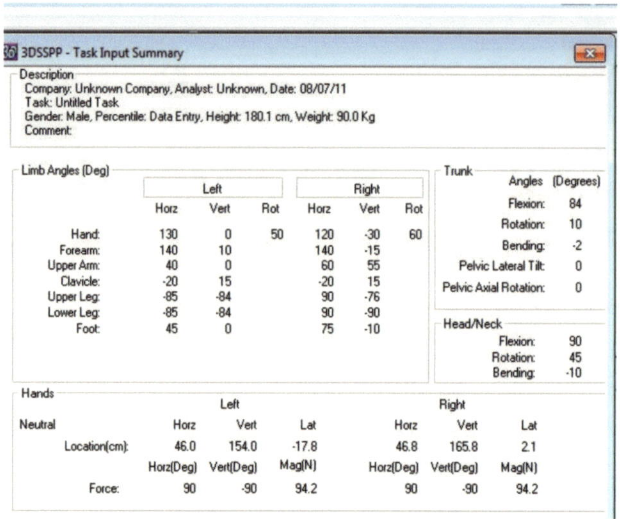

3DSSPP - Task Input Summary

Description
Company: Unknown Company, Analyst: Unknown, Date: 08/07/11
Task: Untitled Task
Gender: Male, Percentile: Data Entry, Height: 180.1 cm, Weight: 90.0 Kg
Comment:

Limb Angles (Deg)

	Left			Right		
	Horz	Vert	Rot	Horz	Vert	Rot
Hand:	130	0	50	120	-30	60
Forearm:	140	10		140	-15	
Upper Arm:	40	0		60	55	
Clavicle:	-20	15		-20	15	
Upper Leg:	-85	-84		90	-76	
Lower Leg:	-85	-84		90	-90	
Foot:	45	0		75	-10	

Trunk

Angles	(Degrees)
Flexion:	84
Rotation:	10
Bending:	-2
Pelvic Lateral Tilt:	0
Pelvic Axial Rotation:	0

Head/Neck

Flexion:	90
Rotation:	45
Bending:	-10

Hands

Neutral	Left			Right		
	Horz	Vert	Lat	Horz	Vert	Lat
Location(cm):	46.0	154.0	-17.8	46.8	165.8	2.1
	Horz(Deg)	Vert(Deg)	Mag(N)	Horz(Deg)	Vert(Deg)	Mag(N)
Force:	90	-90	94.2	90	-90	94.2

Table 7. 3DSSPP Limb Angle Input, Subject 1, Task 1, Pose 2.

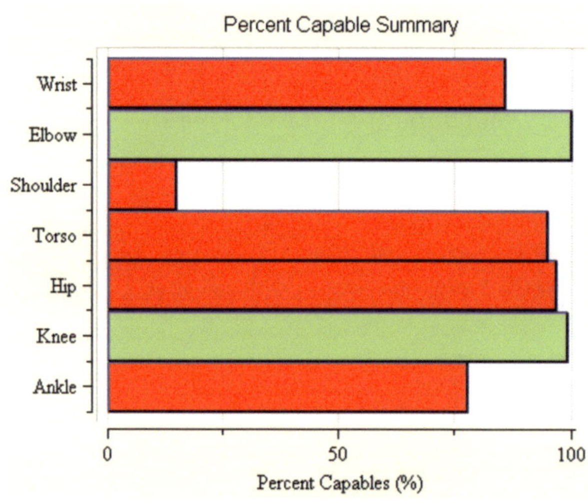

Fig. 22. JACK Output of Percent Capable, Subject 1, Task 1, Pose 2.

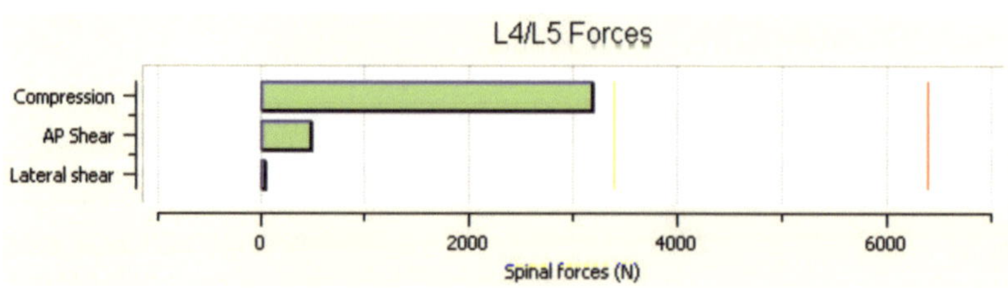

Fig. 23. JACK Output - Forces on L4/L5 for Subject 1, Task 1, Pose 2.

Analysis Recommendations: The low back compression force of 3203.00 is below the NIOSH Back Compression Action Limit of 3400 N, representing a nominal risk of low back injury for most healthy workers.

5.3.1.3 Subject 1, Task 2: Dumping a full waste container

Source Photo JACK 3DSSPP

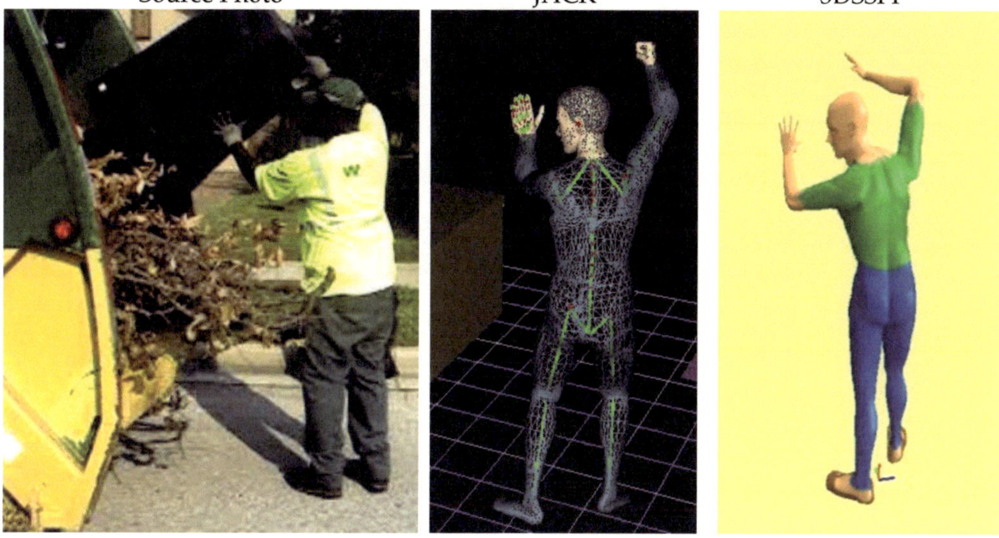

Fig. 24. Dumping Full Waste; JACK, 3DSSPP.

	JACK	3DSSPP
L4/L5 Compression Force (N)	3465	3491

Table 8. Comparison of JACK and 3DSSPP L4/L5 Forces for Subject 1, Task 2 *(See report results in Figures 25 and 28).*

	JACK		3DSSPP
Joint	(% Capable)		(% Capable)
Wrist	98	42	
Elbow	98		90
Shoulder	66	93	
Torso	92		11
Hip	97	99	
Knee	100		70
Ankle	95		

Table 9. Comparison of JACK and 3DSSPP Strength Capability Summary for Subject 1, Task 2 *(See report results in Figures 25 and 27).*

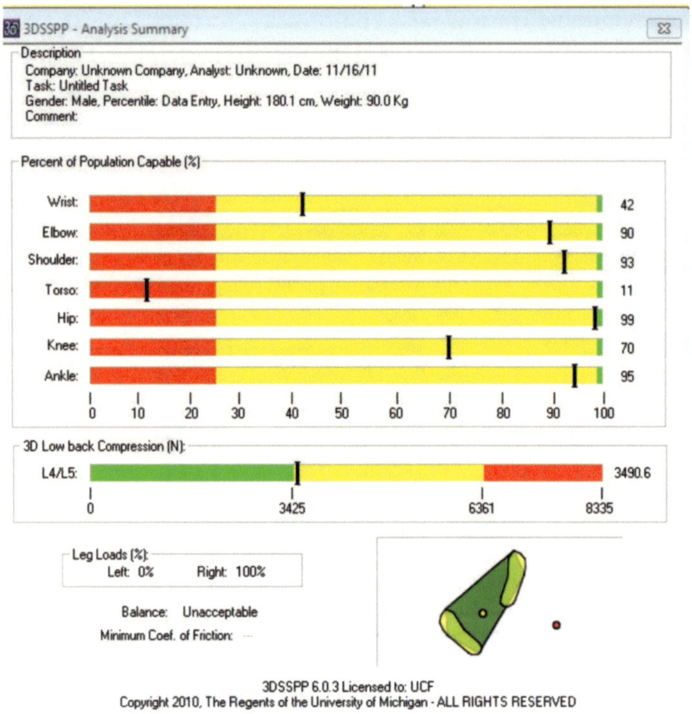

Fig. 25. 3DSSPP Summary Output, Subject 1, Task 2.

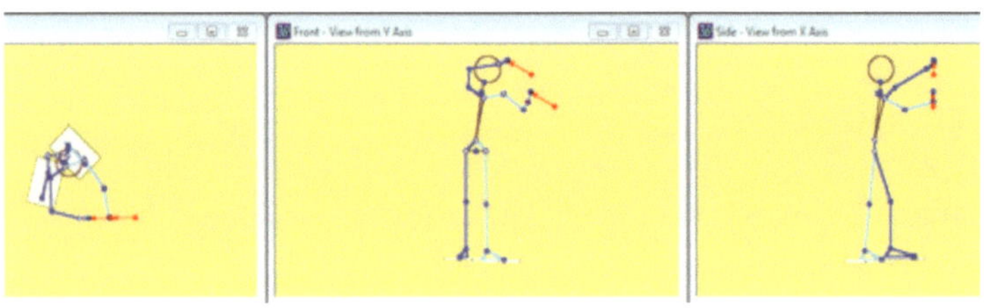

Fig. 26. 3DSSPP Simulation of Limb Angles, Subject 1, Task 2.

5.3.1.3.1 Analysis of Subject 1 - Task 2

The results of both simulations concurred that dumping the waste container was the riskiest task for the waste collection workers due not only to the excessive load but also because of the way the worker is lifting the garbage container. The low back compression force in 3DSSPP and JACK exceeds the NIOSH Back Compression Design Limit of 3400N. Workers should avoid twisting while dumping the waste container to avoid awkward postures of the body joints.

The JACK low back analysis report suggests the following ways to reduce the back compressive forces:

1. Reducing the weight of the load.
2. Changing the job environment such that the worker does not need to stoop to lift the load (avoid having to bend over).

3. Ensuring the load is small, such that it can be held close to the body.
4. Avoiding asymmetric (twisted) postures.

3DSSPP - Task Input Summary

Description
Company: Unknown Company, Analyst: Unknown, Date: 11/16/11
Task: Untitled Task
Gender: Male, Percentile: Data Entry, Height: 180.1 cm, Weight: 90.0 Kg
Comment:

Limb Angles (Deg)

	Left Horz	Left Vert	Left Rot	Right Horz	Right Vert	Right Rot
Hand:	0	75	-60	180	19	-20
Forearm:	80	15		167	9	
Upper Arm:	55	-25		92	30	
Clavicle:	10	15		0	25	
Upper Leg:	-85	-84		90	-71	
Lower Leg:	-85	-84		90	-90	
Foot:	45	0		75	0	

Trunk

	Angles (Degrees)
Flexion:	84
Rotation:	30
Bending:	-7
Pelvic Lateral Tilt:	0
Pelvic Axial Rotation:	0

Head/Neck

Flexion:	90
Rotation:	50
Bending:	-10

Hands

Neutral	Left Horz	Left Vert	Left Lat	Right Horz	Right Vert	Right Lat
Location(cm):	51.2	148.3	-49.4	51.6	176.1	-28.4
	Horz(Deg)	Vert(Deg)	Mag(N)	Horz(Deg)	Vert(Deg)	Mag(N)
Force:	0	29	94.2	0	29	94.2

Table 10. 3DSSPP Limb Angle Input, Subject 1, Task 2.

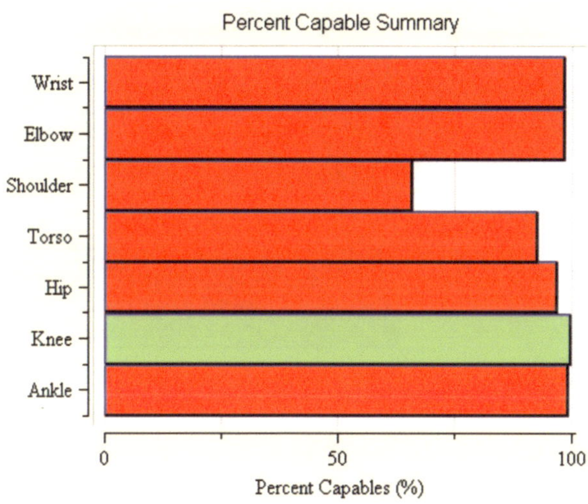

Fig. 27. JACK Output of Percent Capable, Subject 1, Task 2.

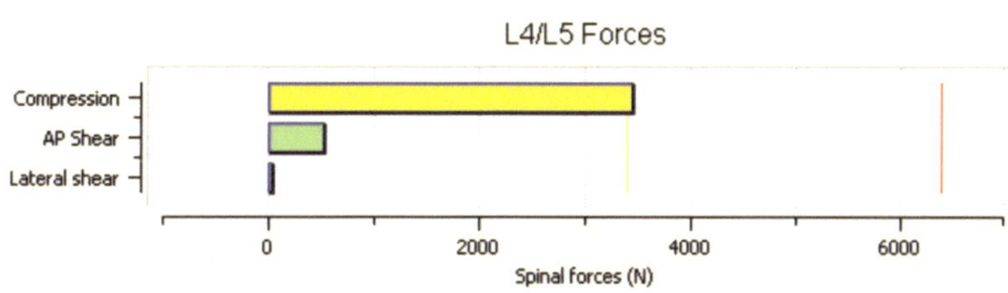

Fig. 28. JACK Output - Forces on L4/L5 for Subject 1, Task 2.

The percent of the population capable of performing this posture ranges from 11-99% according to 3DSSPP. The torso area exhibits the most strain; 11% only of population is capable of performing this task, this percent falls below the NIOSH Upper Limit Value. On the other hand, JACK indicated that 92% of the population will be able to perform this task with respect to the torso joint. As per the knee joint 100 % of the population will be able to perform this pose, while 3DSSPP indicated that 70% of population will perform this pose. Also, it was noticed that there is a significant difference between both packages for the wrist joint; JACK indicates that 98% of population will perform this pose while according to 3DSSPP only 42% of the population will be able to perform the dumping task. As was mentioned previously, this difference is due the degrees of freedom in manipulating the joints in 3DSSPP. For the other joints, both packages indicated that they would fall within the yellow zone.

5.3.2 Subject 2: Disaster management project

5.3.2.1 Subject 2, Task 1, Pose 1: Victim extraction

LA Search and Rescue pull woman out of rubble, 12 Jan 2010 Haiti Quake, Virtual Environment JACK, 3DSSPP.

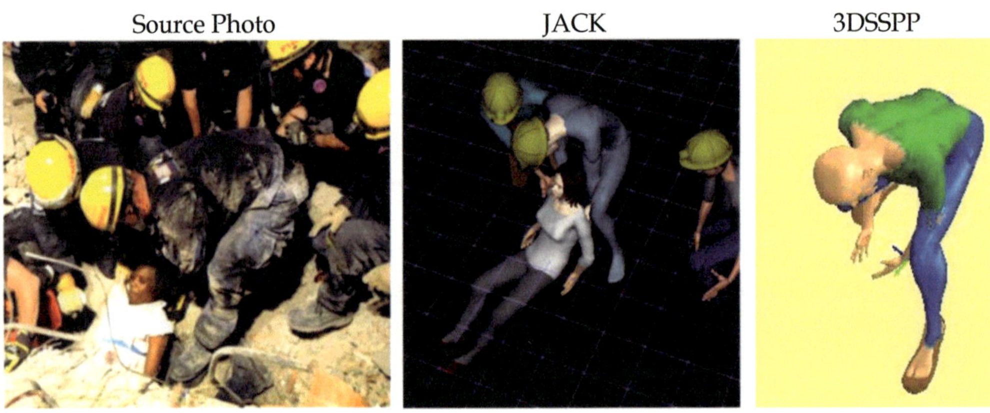

Fig. 29. LA Search and Rescue pull woman out of rubble, 12 Jan 2010 Haiti Quake.

	JACK	3DSSPP
L4/L5 Compression Force (N)	6658.	6715.

Table 11. Comparison of JACK and 3DSSPP L4/L5 Forces for Subject 2, Task 1 *(See report results in Figure 30 and Table 14).*

	JACK	3DSSPP
Joint	(% Capable)	(% Capable)
Wrist	13 14	
Elbow	74	85

	JACK		3DSSPP
Shoulder	53	99	
Torso	80		77
Hip	96	44	
Knee	86		25
Ankle	99	93	

Table 12. Comparison of JACK and 3DSSPP Strength Capability Summary for Subject 2, Task 1*(See report results in Figures 30 and 32).*

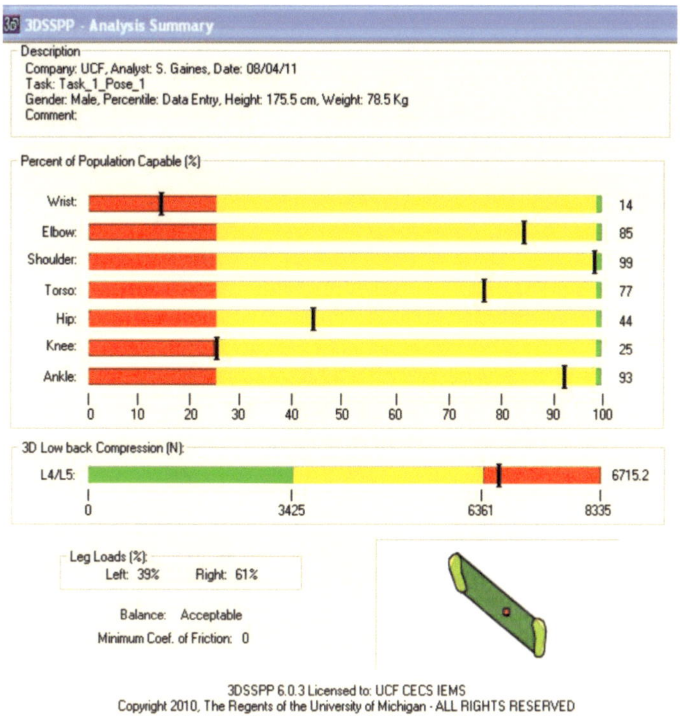

Fig. 30. 3DSSPP Summary Output, Subject 2, Task 1.

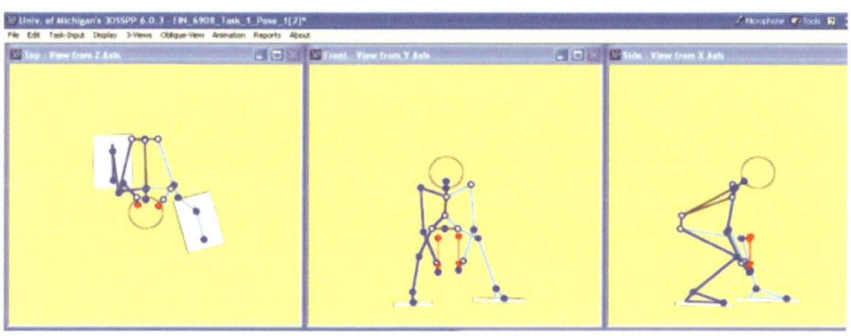

Fig. 31. 3DSSPP Simulation of Limb Angles, Subject 2, Task 1.

3DSSPP - Task Input Summary

Description
Company: UCF, Analyst: S. Gaines, Date: 08/04/11
Task: Task_1_Pose_1
Gender: Male, Percentile: Data Entry, Height: 175.5 cm, Weight: 78.5 Kg
Comment:

Limb Angles (Deg)

	Left Horz	Left Vert	Left Rot	Right Horz	Right Vert	Right Rot
Hand:	124	-50	10	100	-50	20
Forearm:	124	-61		131	-59	
Upper Arm:	-52	-84		-109	-79	
Clavicle:	-20	15		-20	15	
Upper Leg:	71	-8		77	-27	
Lower Leg:	39	-66		-82	-39	
Foot:	75	-1		91	-2	

Trunk

Angles	(Degrees)
Flexion:	38
Rotation:	5
Bending:	-1
Pelvic Lateral Tilt:	0
Pelvic Axial Rotation:	0

Head/Neck

Flexion:	95
Rotation:	20
Bending:	0

Hands

Neutral	Left Horz	Left Vert	Left Lat	Right Horz	Right Vert	Right Lat
Location(cm):	18.0	25.2	-2.9	17.2	23.7	11.2
	Horz(Deg)	Vert(Deg)	Mag(N)	Horz(Deg)	Vert(Deg)	Mag(N)
Force:	90	-90	244.7	90	-90	244.7

Table 13. 3DSSPP Limb Angle Input, Subject 2, Task 1.

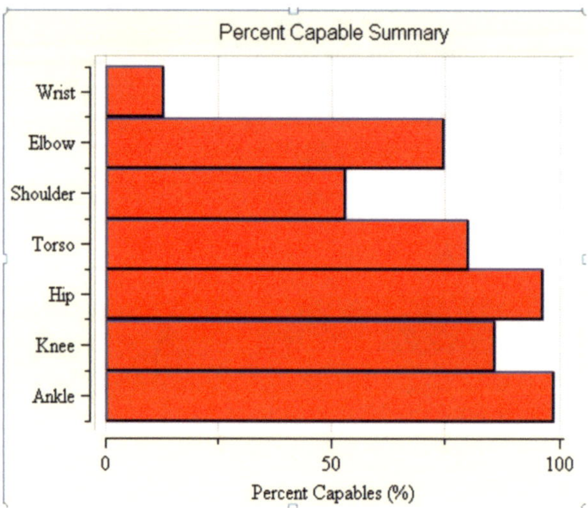

Fig. 32. JACK Output of Percent Capable, Subject 2, Task 1.

TSB Ergonomic Report

Time	Task						
0	Primary_Pose_1						
	L4/L5 Forces (N)			L4/L5 Moments (N)			
	COMPRESSION	AP SHEAR	LATERAL SHEAR	L4/L5 X	L4/L5 Y	L4/L5 Z	
0.967	6658.387	1350.471	-19.447	318.393	24.353	-32.577	

Table 14. JACK TSB Ergonomic Report - Forces on L4/L5, Subject 2, Task 1.

5.3.2.1.1 Analysis of Subject 2 - Task 1, Pose 1

The compression forces on L4/L5 are similar between JACK and 3DSSPP, that is, they are within 57N of each other, or less than a 1% difference. For the percent capable, both packages predicted 7 of the 8 joints would fall within the caution, or yellow zone and one joint would fall within the red zone. These tasks appear to have been simulated in a similar fashion and the output is comparable.

For this task either software package would generate similar results. In both cases, the forces on L4/L5 fall well above the NIOSH recommended upper limit of 3400N, meaning performance of this task, especially if repeated frequently using these postures and loads will likely result in injury to the first responder. In fact, both of the software packages calculated a force on L4/L5 of greater than the maximum limit allowed by NIOSH of 6400N. This task should not be performed by only one person. At these angles and loads, at least two people must assist in lifting the load.

Analysis Recommendations: The low back compression force of 3465.00 is above the NIOSH Back Compression Action Limit of 3400 N, representing an increased risk of low back injury for some workers. It is recommended that this job analyzed further for ways to reduce low back forces.

5.3.2.2 Subject 2, Task 2, Pose 5: Moving the injured

Source Photo JACK 3DSSPP

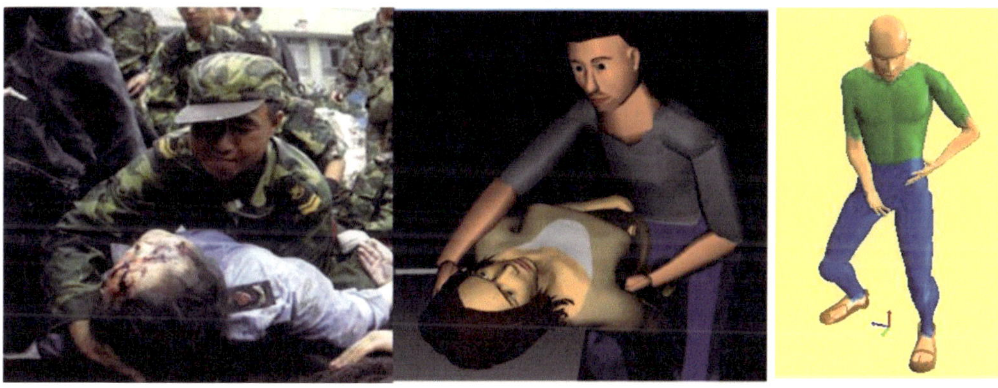

Fig. 33. Rescuers carry injured quake victim from collapsed building in Beichuan County, China. May, 2008.

	JACK	3DSSPP
L4/L5 Compression Force (N)	6853.	6966.

Table 15. Comparison of JACK and 3DSSPP L4/L5 Forces for Subject 2, Task 2 *(See report results in Figure 34 and Table 18).*

Joint	(% Capable)	(% Capable)
Wrist	3	0

Joint	(% Capable)	(% Capable)
Elbow	50	0
Shoulder	0	0
Torso	48	20
Hip	90	3
Knee	99	98
Ankle	87	2

Table 16. Comparison of JACK and 3DSSPP Strength Capability Summary for Subject 2, Task 2 *(See report results in Figures 34 and 36).*

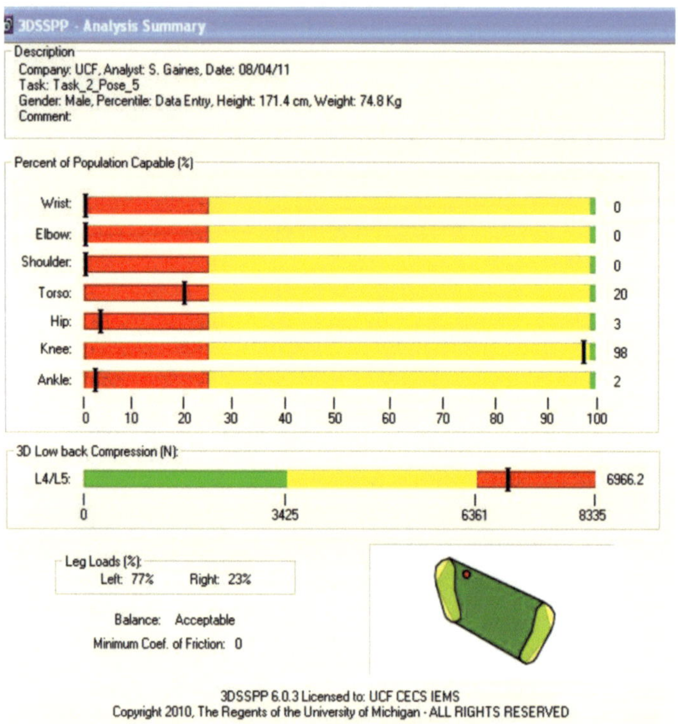

Fig. 34. 3DSSPP Summary Output, Subject 2, Task 2.

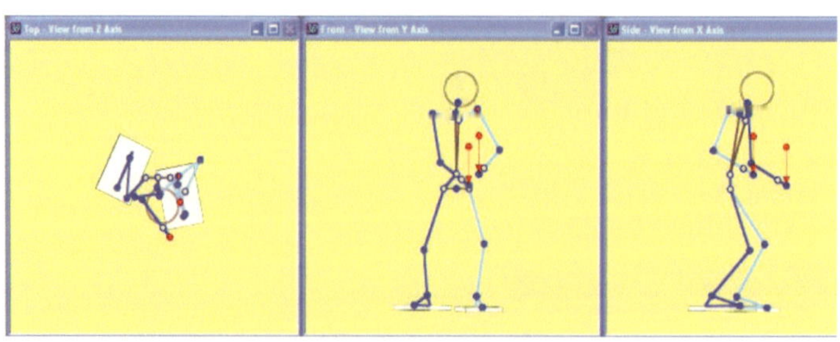

Fig. 35. 3DSSPP Simulation of Limb Angles, Subject 2, Task 2.

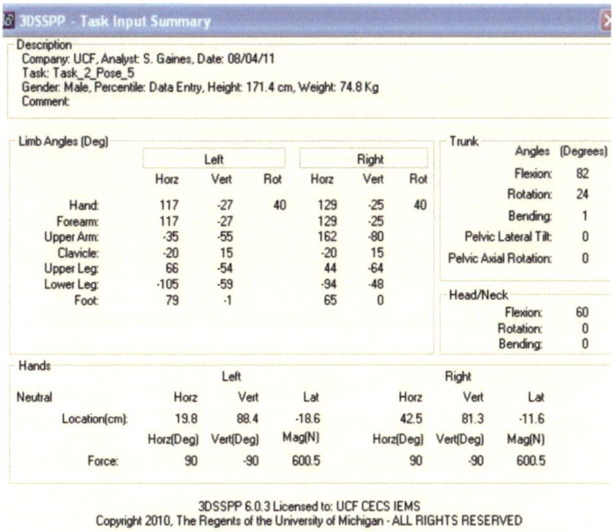

3DSSPP - Task Input Summary

Description
Company: UCF, Analyst: S. Gaines, Date: 08/04/11
Task: Task_2_Pose_5
Gender: Male, Percentile: Data Entry, Height: 171.4 cm, Weight: 74.8 Kg
Comment:

Limb Angles (Deg)

	Left			Right		
	Horz	Vert	Rot	Horz	Vert	Rot
Hand:	117	-27	40	129	-25	40
Forearm:	117	-27		129	-25	
Upper Arm:	-35	-55		162	-80	
Clavicle:	-20	15		-20	15	
Upper Leg:	66	-54		44	-64	
Lower Leg:	-105	-59		-94	-48	
Foot:	79	-1		65	0	

Trunk

Angles	(Degrees)
Flexion:	82
Rotation:	24
Bending:	1
Pelvic Lateral Tilt:	0
Pelvic Axial Rotation:	0

Head/Neck

Flexion:	60
Rotation:	0
Bending:	0

Hands

Neutral	Left			Right		
	Horz	Vert	Lat	Horz	Vert	Lat
Location(cm):	19.8	88.4	-18.6	42.5	81.3	-11.6
	Horz(Deg)	Vert(Deg)	Mag(N)	Horz(Deg)	Vert(Deg)	Mag(N)
Force:	90	-90	600.5	90	-90	600.5

3DSSPP 6.0.3 Licensed to: UCF CECS IEMS
Copyright 2010, The Regents of the University of Michigan - ALL RIGHTS RESERVED

Table 17. 3DSSPP Limb Angle Input, Subject 2, Task 2.

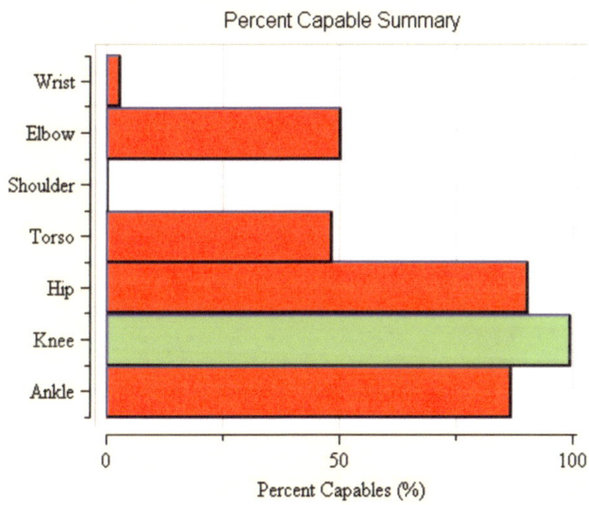

Fig. 36. JACK Output of Percent Capable, Subject 2, Task 2.

TSB Ergonomic Report							
Time	Task						
0	Primary_Pose_1						
		L4/L5 Forces (N)			L4/L5 Moments (N)		
		COMPRESSION	AP SHEAR	LATERAL SHEAR	L4/L5 X	L4/L5 Y	L4/L5 Z
5		6853.777	1145.615	-69.81	326.016	36.596	44.323

Table 18. JACK TSB Ergonomic Report - Forces on L4/L5, Subject 2, Task 2.

5.3.2.2.1 Analysis of Subject 2 - Task 2, Pose 5

The forces on L4/L5 generated by the two software packages are similar, with results within 2% of each other. The compression force on L4/L5 calculated by both packages indicates that this task activity is above the maximum allowable NIOSH limit of 6400N. Essentially this task should not be performed by one person and not in the postures exhibited.

For the percent capable, 3DSSPP predicted this is a more difficult task for the majority of the population to perform than JACK. In fact, 3DSSPP calculated that 5% or less of the population could perform the task for 5 of the 8 joints analyzed. A review of the overall data found in Figures 34 and 36, indicates that the shoulder, for example, has a 98-100% capable in all areas except one. That one is noted to be at 0%, so the software automatically accepts the lowest number. The other significant difference in was seen in the hip joint. 3DSSPP said only 3% of the population could perform the task, while JACK thinks 90% of the population can perform the task. Manually manipulating the postures in 3DSSPP was the only way to visually achieve the "same" posture. 3DSSPP seems to have less ability to gradually change the postures. When the center of hips is moved for example, all of the other angles changed dramatically.

For this task, JACK appears to be a better simulation tool. It allows more detailed manipulation of hand postures and also gives more flexibility with regard to torso rotations and flexibility. The greater ability to specify the angles of shoulder rotation, elevation, and lift are also pivotal in this task analysis. See Figure 37 for an example of how JACK allows this detailed input. The force on the shoulder was probably one of the greatest, other than on L4/L5 for this task.

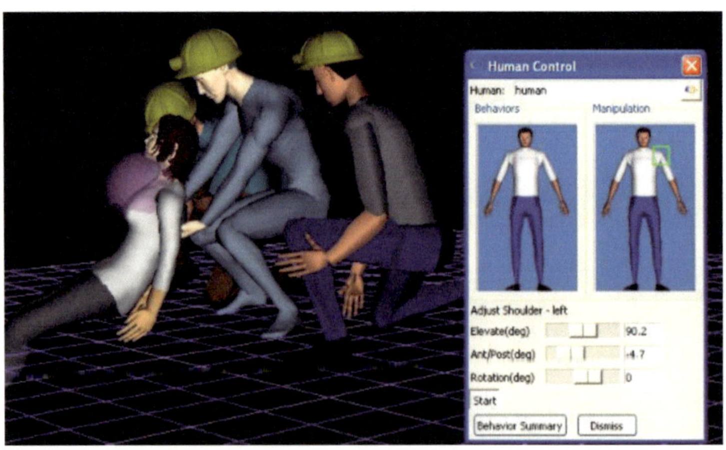

Fig. 37. The Human Control tab in JACK allows greater manipulation of the shoulder joint, pivotal in this task.

5.3.2.3 Subject 2, Task 3, Pose 4: Supply distribution

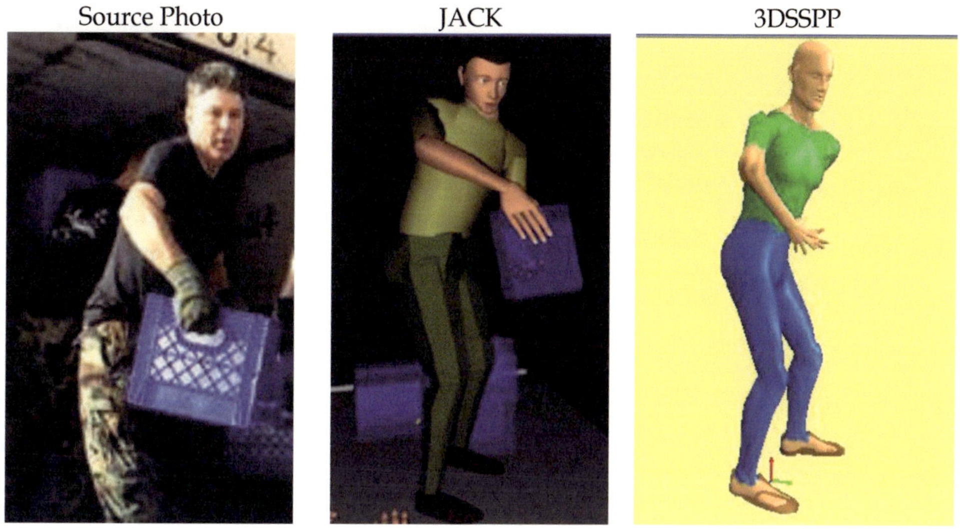

Fig. 38. New Jersey National, Guard's Response to Hurricane Katrina
Photo courtesy of pdcbank.state.nj.us

	JACK	3DSSPP
L4/L5 Compression Force (N)	2626.	2461.

Table 19. Comparison of JACK and 3DSSPP L4/L5 Forces for Subject 2, Task 3 *(See report results in Figure 39 and Table 22).*

Joint	JACK (% Capable)		3DSSPP (% Capable)
Wrist	99	67	
Elbow	100		98
Shoulder	99	99	
Torso	98		96
Hip	98	91	
Knee	100		99
Ankle	97	87	

Table 20. Comparison of JACK and 3DSSPP Strength Capability Summary for Subject 2, Task 3. *(See report results in Figures 39 and 41).*

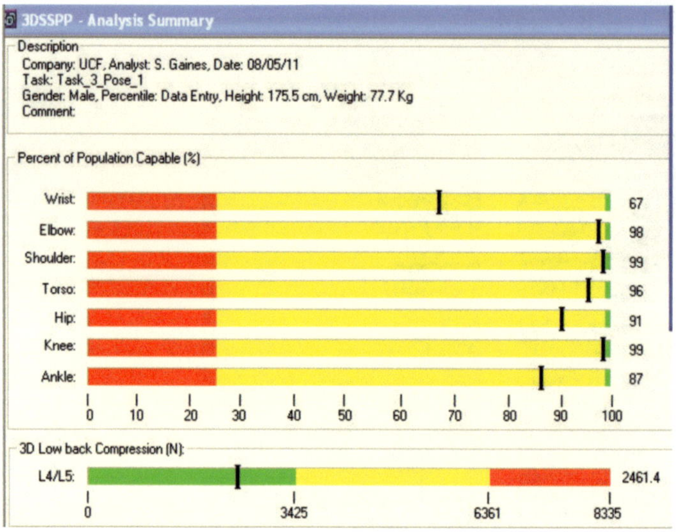

Fig. 39. 3DSSPP Summary Output, Subject 2, Task 3.

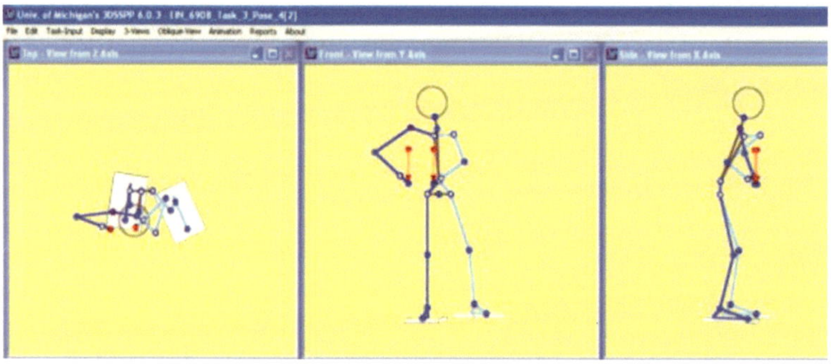

Fig. 40. 3DSSPP Simulation of Limb Angles, Subject 2, Task 3.

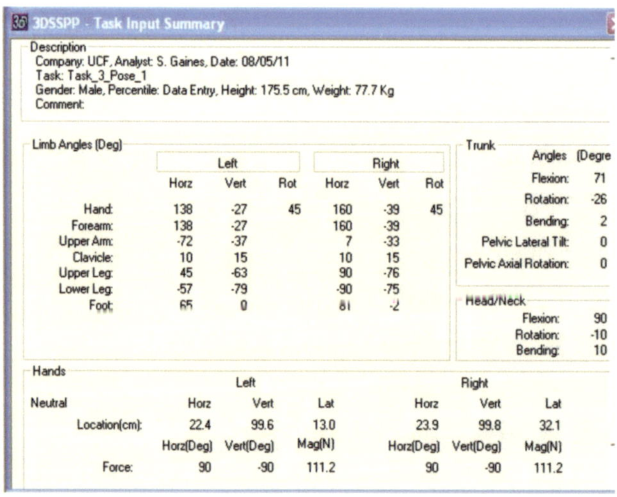

3DSSPP - Task Input Summary

Description
Company: UCF, Analyst: S. Gaines, Date: 08/05/11
Task: Task_3_Pose_1
Gender: Male, Percentile: Data Entry, Height: 175.5 cm, Weight: 77.7 Kg
Comment:

Limb Angles (Deg)

	Left			Right		
	Horz	Vert	Rot	Horz	Vert	Rot
Hand:	138	-27	45	160	-39	45
Forearm:	138	-27		160	-39	
Upper Arm:	-72	-37		7	-33	
Clavicle:	10	15		10	15	
Upper Leg:	45	-63		90	-76	
Lower Leg:	-57	-79		-90	-75	
Foot:	65	0		81	-2	

Trunk

	Angles (Degre
Flexion:	71
Rotation:	-26
Bending:	2
Pelvic Lateral Tilt:	0
Pelvic Axial Rotation:	0

Head/Neck

Flexion:	90
Rotation:	-10
Bending:	10

Hands

Neutral	Left			Right		
	Horz	Vert	Lat	Horz	Vert	Lat
Location(cm):	22.4	99.6	13.0	23.9	99.8	32.1
	Horz(Deg)	Vert(Deg)	Mag(N)	Horz(Deg)	Vert(Deg)	Mag(N)
Force:	90	-90	111.2	90	-90	111.2

Table 21. 3DSSPP Limb Angle Input, Subject 2, Task 3.

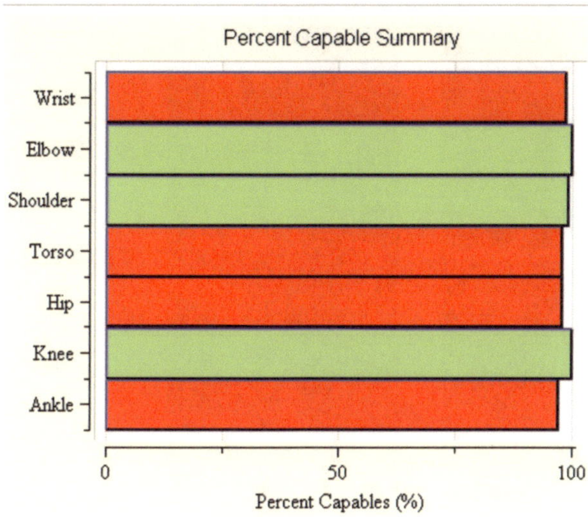

Fig. 41. JACK Output of Percent Capable, Subject 2, Task 3.

TSB Ergonomic Report							
Time	Task						
0	Primary_Pose_1						
		L4/L5 Forces (N)			L4/L5 Moments (N)		
		COMPRESSION	AP SHEAR	LATERAL SHEAR	L4/L5 X	L4/L5 Y	L4/L5 Z
5		2626.195	381.153	-46.992	105.28	11.412	28.32

Table 22. JACK TSB Ergonomic Report - Forces on L4/L5, Subject 2, Task 3.

5.3.2.3.1 Analysis of Subject 2 - Task 3, Pose 4

Both JACK and 3DSSPP generated similar L4/L5 compression force calculations, and while slightly different, the forces on L4/L5 for this task fell below the NIOSH recommended upper limit of 3400N for both packages. The frequency with which this task may be repeated was not considered, and would inevitably generate a fatigue factor if, for example, an entire truckload of supplies at this weight using this posture were unloaded. Both software packages calculated that the percent of the population which could perform this task was over 90% for every joint except one (3DSSPP said only 67% of the population could perform these wrist manipulations). Either package would be able to adequately simulate this task. 5.3.3 SUBJECT 3: INTERACTIVE GAMING PROJECT

5.3.3.1 Subject 3, Task 1: Overhead Catch

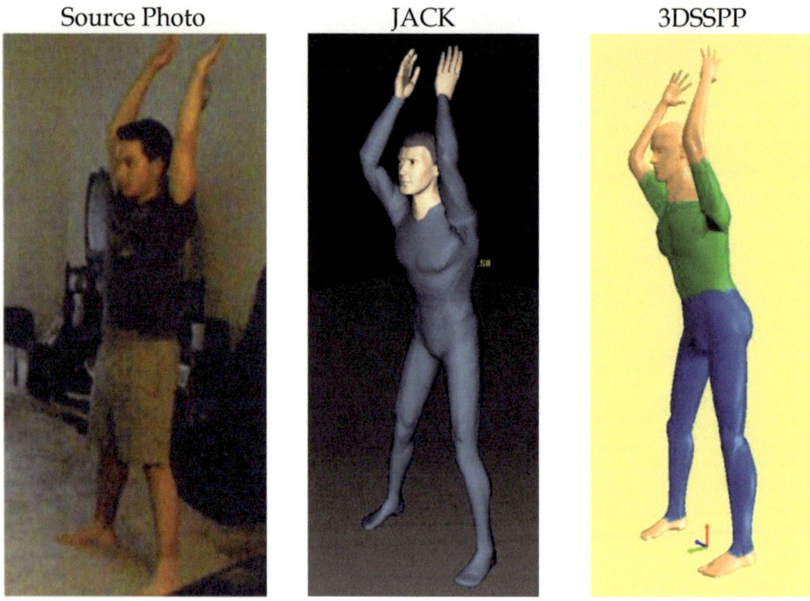

Fig. 42. Overhead Catch; JACK; 3DSSPP.

	JACK	3DSSPP
L4/L5 Compression Force (N)	942	1074

Table 23. Comparison of JACK and 3DSSPP L4/L5 Forces for Subject 3, Task 1 *(See report results in Figures 43 and 47).*

	JACK	3DSSPP
Joint	(% Capable)	(% Capable)
Wrist	100	99
Elbow	100	100
Shoulder	100	99
Torso	100	99
Hip	99	98
Knee	100	99
Ankle	100	99

Table 24. Comparison of JACK and 3DSSPP Strength Capability Summary for Subject 3, Task 1. *(See report results in Figures 43 and 46).*

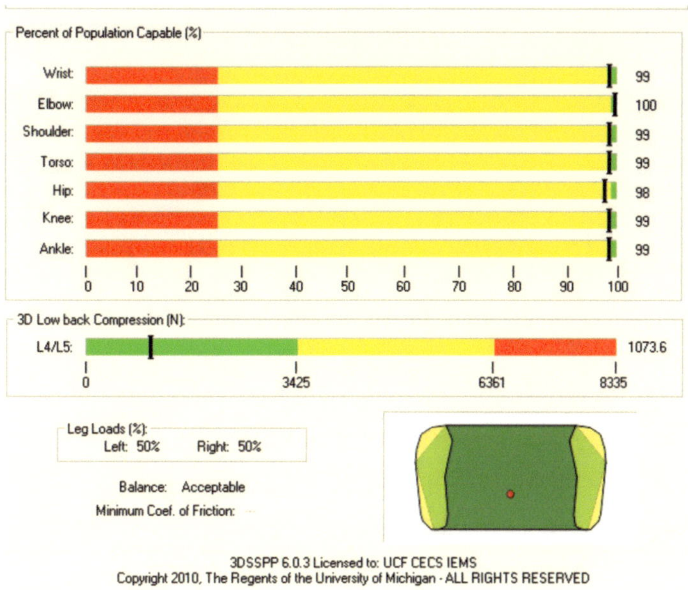

Fig. 43. 3DSSPP Summary Output, Subject 3, Task 1.

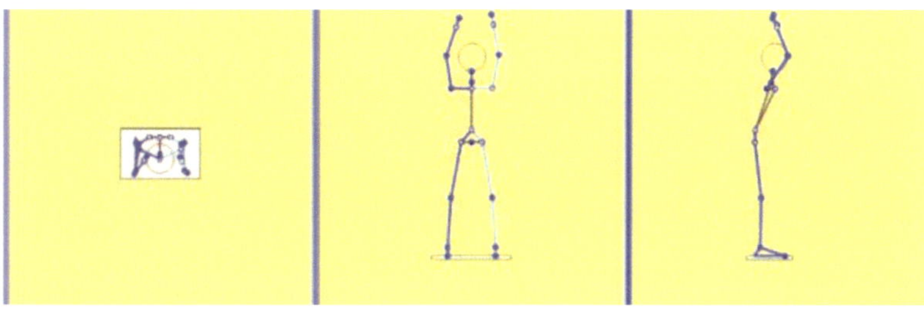

Fig. 44. 3DSSPP Simulation of Limb Angles, Subject 3, Task 1.

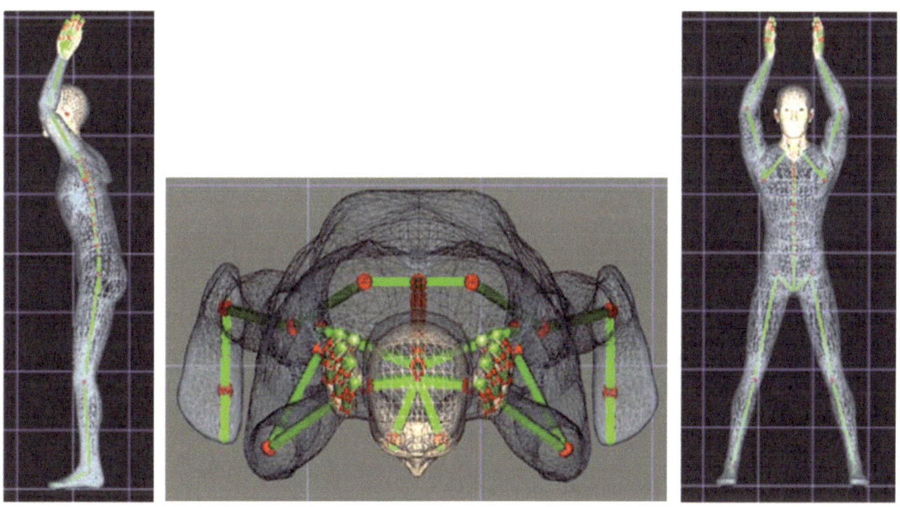

Fig. 45. JACK Simulation, Skeletal View aids in limb angle calculations.

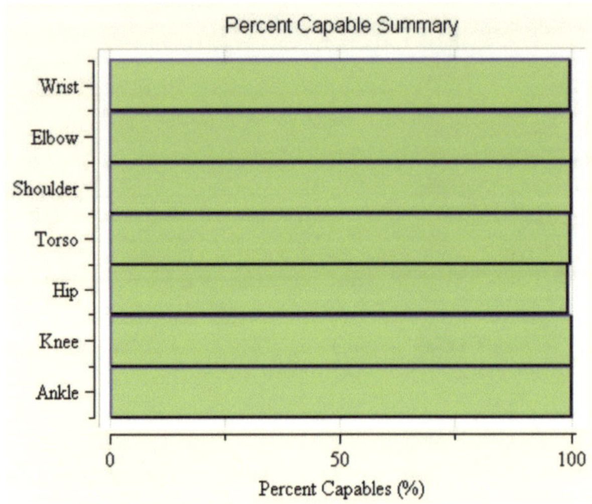

Fig. 46. JACK Output of Percent Capable, Subject 3, Task 1.

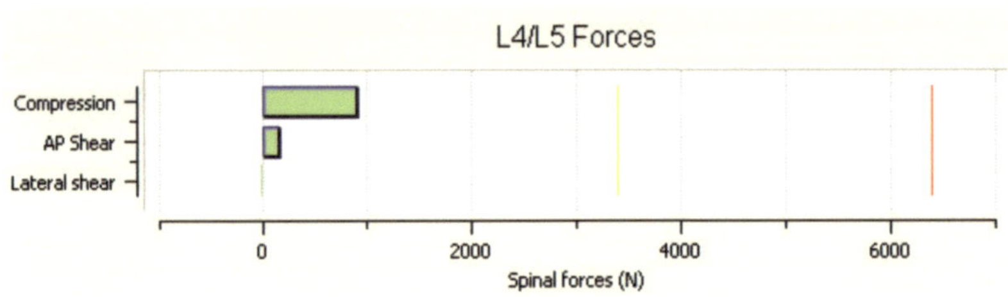

Fig. 47. JACK Output - Forces on L4/L5 for Subject 3, Task 1.

5.3.3.1.1 Analysis of Subject 3 - Task 1

For Task 2, the compression force in the lower back is below the NIOSH Back Compression Action Limit of 3400 N. JACK reports this force is 942 N and 3DSSPP calculates a force of 1074N. Both of these values designate this task as low risk for an average person. The percent of the population capable of performing this posture ranges from 98 to 100 percent by JACK calculations and 98 to 100 percent according to 3DSSPP. According to 3DSSPP, the wrist, shoulder, torso, hip, knee and ankle area exhibit the most strain and are slightly past the NIOSH Strength Design Limit (SDL) value. JACK did not perceive any warning, as all of the joints remained in the "green" zone and were within the SDL. The lowest percentage of 98 is still generally high and may be determined a tolerable risk. However, note should be made of the yellow designation that expose joints of the body where the user population may experience some limitations in performing this task. In addition, it is important to note that although this posture is within an acceptable level according to NIOSH standards, that this posture involves holding the hands above the head with hands held behind the head. This sort of posture can lead to increased heart rate should be avoided on a repetitive or prolonged basis.

Analysis Recommendations: The low back compression force of 924.00 is below the NIOSH Back Compression Action Limit of 3400 N, representing a nominal risk of low back injury for most healthy workers.

5.3.3.2 Subject 3, Task 2: Low-ball upper-limb save

Source Photo JACK 3DSSPP

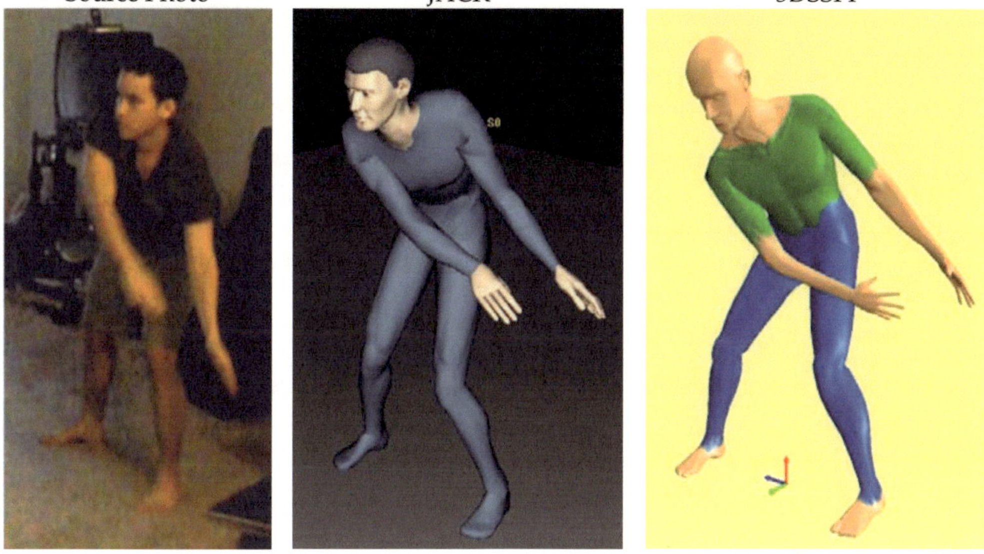

Fig. 48. Low-Ball, Upper-Limb Save; JACK; 3DSSPP.

	JACK	3DSSPP
L4/L5 Compression Force (N)	1942	2071

Table 25. Comparison of JACK and 3DSSPP L4/L5 Forces for Subject 3, Task 2 *(See report results in Figures 49 and 53).*

	JACK	3DSSPP
Joint	(% Capable)	(% Capable)
Wrist	100	99
Elbow	100	100
Shoulder	100	99
Torso	99	98
Hip	99 95	
Knee	100	99
Ankle	100	99

Table 26. Comparison of JACK and 3DSSPP Strength Capability Summary for Subject 3, Task 2. *(See report results in Figures 49 and 52).*

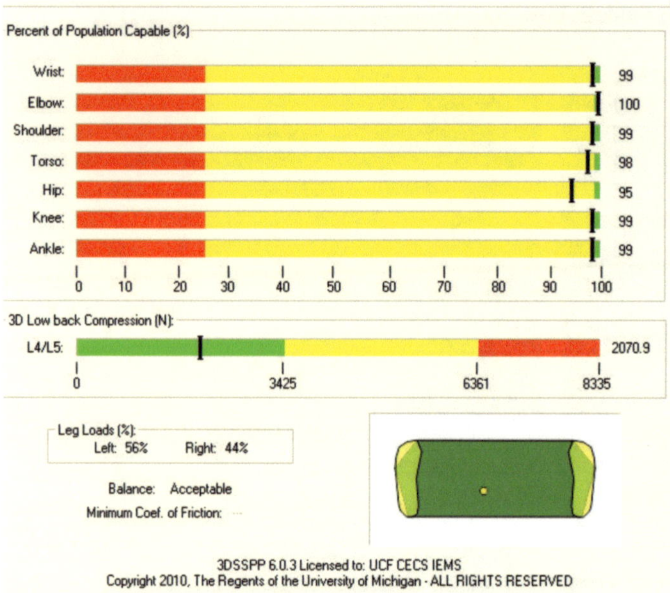

Fig. 49. 3DSSPP Summary Output, Subject 3, Task 2.

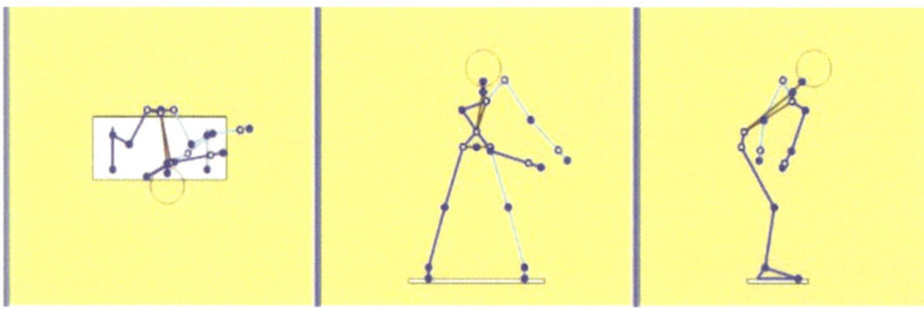

Fig. 50. 3DSSPP Simulation of Limb Angles, Subject 3, Task 2.

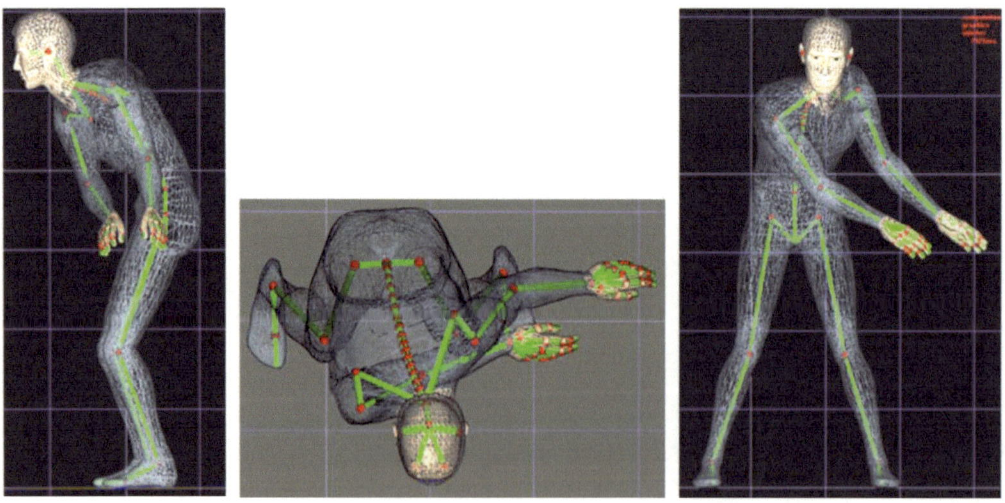

Fig. 51. JACK Simulation, Skeletal View aids in limb angle calculations.

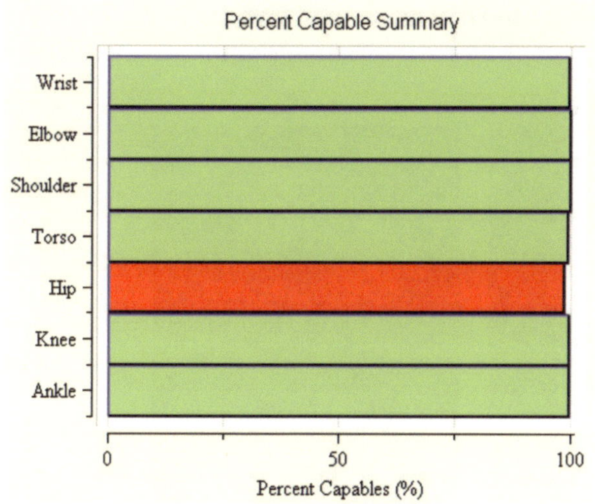

Fig. 52. JACK Output of Percent Capable, Subject 3, Task 2.

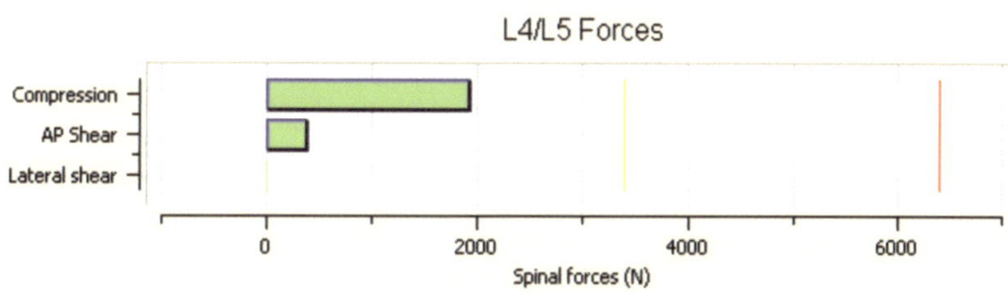

Fig. 53. JACK Output - Forces on L4/L5 for Subject 3, Task 2.

5.3.3.2.1 Analysis of Subject 3 - Task 2

For Task 3, the compression force in the lower back (L4/L5) is below the NIOSH Back Compression Action Limit of 3400 N. JACK reports this force is 1942 N and 3DSSPP calculates a force of 2071 N. Both of these values designate this task as low risk for an average person. The percent of the population capable of performing this posture ranges from 99 to 100 percent (JACK) and 95 to 100 percent (3DSSPP). According to JACK, the torso and hip area exhibit the most strain and are slightly past the NIOSH Strength Design Limit (SDL) value. 3DSSPP predicted that in addition to the torso and hip, the wrist, shoulder, knee, and ankle also exceed the action limit value. Although the lowest percentage of 95 is high and may be generally regarded as tolerable risk, the yellow designation shows the areas of the body that may limit the average person from performing this task safely. The twisting of the torso is a risky posture and reaching across the body's centerline contributes to the compression of the back, as well as the variation of load on each leg. Repetitively performing this posture may increase risk of WMSD.

Analysis Recommendations: The low back compression force of 1942.00 is below the NIOSH Back Compression Action Limit of 3400 N, representing a nominal risk of low back injury for most healthy workers.

5.3.3.3 Subject 3, Task 3: Low-ball lower-limb save

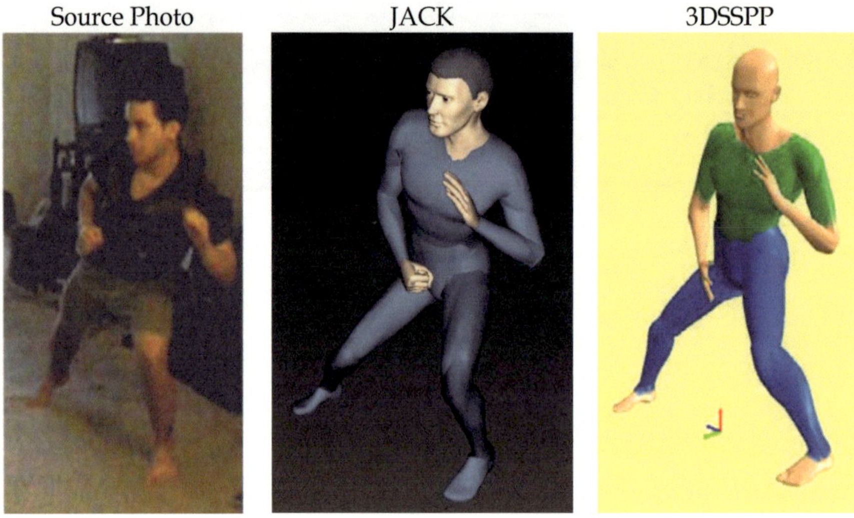

Fig. 54. Low-Ball Lower-Limb Save; JACK; 3DSSPP.

	JACK	3DSSPP
L4/L5 Compression Force (N)	1656	1638

Table 27. Comparison of JACK and 3DSSPP L4/L5 Forces for Subject 3, Task 3 *(See report results in Figures 55 and 59).*

	JACK	3DSSPP
Joint	(% Capable)	(% Capable)
Wrist	100	99
Elbow	100	100
Shoulder	100	99
Torso	100	99
Hip	99 98	
Knee	79	98
Ankle	100	99

Table 28. Comparison of JACK and 3DSSPP Strength Capability Summary for Subject 3, Task 3. *(See report results in Figures 55 and 58).*

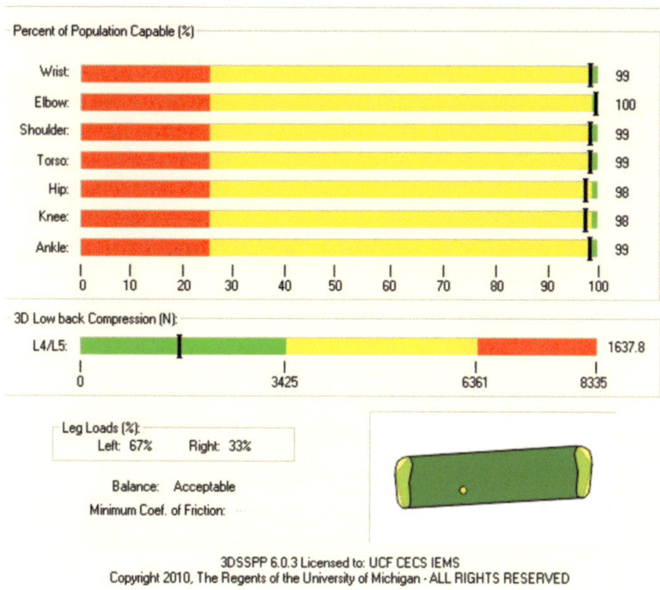

Fig. 55. 3DSSPP Summary Output, Subject 3, Task 3.

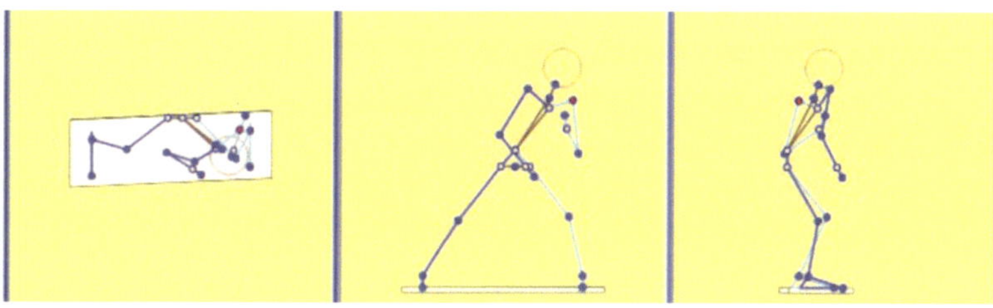

Fig. 56. 3DSSPP Simulation of Limb Angles, Subject 3, Task 3.

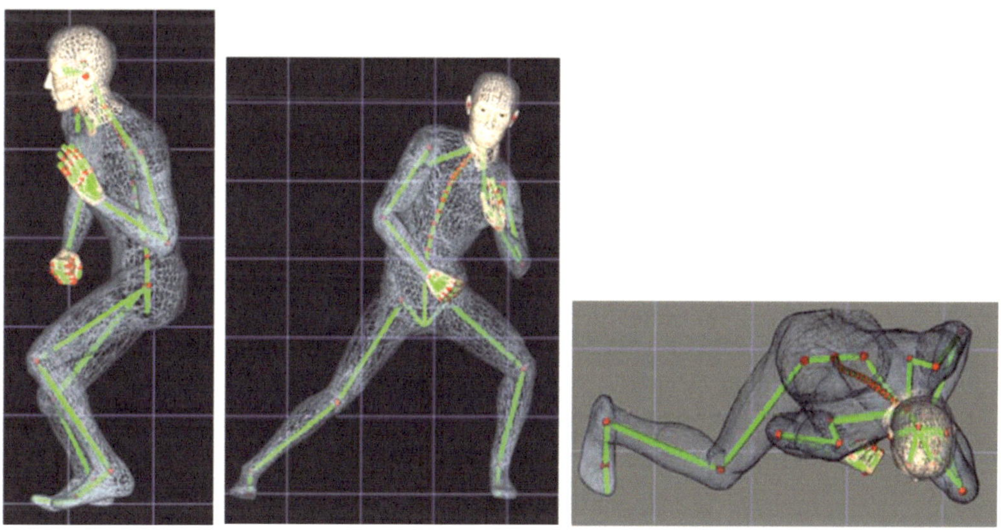

Fig. 57. JACK Simulation, Skeletal View aids in limb angle calculations.

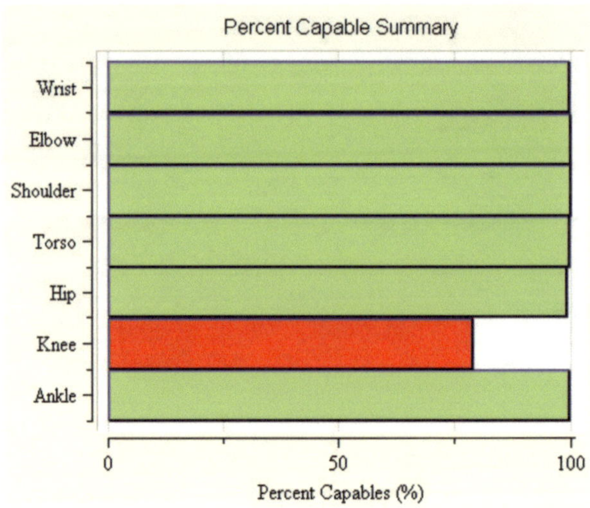

Fig. 58. JACK Output of Percent Capable, Subject 3, Task 3.

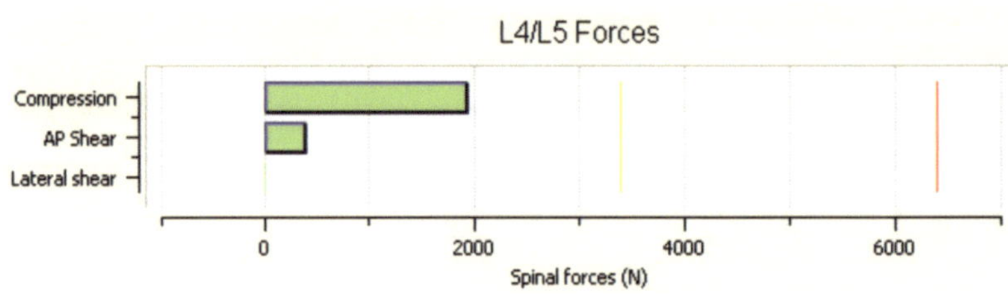

Fig. 59. JACK Output - Forces on L4/L5 for Subject 3, Task 3.

5.3.3.3.1 Analysis of Subject 3 - Task 3

For Task 4, the compression force in the lower back (L4/L5) is below the NIOSH Back Compression Action Limit of 3400 N. JACK reports this force is 1656 N and 3DSSPP calculates a force of 1638 N. Both of these values designate this task as low risk for an average person. The percentage of the population capable of performing this posture ranges from 79 to 100 percent. The hip and knee areas exhibit the most strain based on both JACK and 3DSSPP analysis. According to JACK, only 79% of the population of males can perform this posture with the load that is placed at the joint of the knee. This value is past the NIOSH design limit. This move may be very risky for users that perform this posture during game play on a regular basis. 3DSSPP predicted that the wrist, shoulder, knee, and ankle also exceed the action limit value. Here, the center or gravity is transferred more to the left side, which creates a variation of load on each leg. Repetitively performing this posture could lead to loss of balance and increase risk of falling or overuse injuries.

In evaluating the Kinect gaming tasks, both software analyses identify all poses as generally safe to perform. The biomechanical load on the joints may be underestimated since the jumping, rapid acceleration and deceleration of body segments, and the duration and frequency of movement are not considered in a static strength prediction. Considering these additional factors, the tasks may show different user capabilities. Thus, for the application of highly repetitive tasks, with short duration, and higher velocity of movements, these evaluation tools are limited in assessing risks. However, both programs do acknowledge this limitation as they are based on static strength predictions and employ NIOSH guidelines. Also, it is important to note that the two software tools are developed for evaluating occupational tasks. This data is useful in the preliminary assessment of the postures involved in game play, but may not be conclusive. The JACK software includes Rapid Upper Limb Assessment Tool (RULA), NIOSH, Metabolic Energy Expenditure, and Fatigue Analysis tools may be useful in further evaluating the impact of the task on the user. These features may be further explored in later experimentation.

Analysis Recommendations: The low back compression force of 1656.00 is below the NIOSH Back Compression Action Limit of 3400 N, representing a nominal risk of low back injury for most healthy workers.

6. Conclusion

The intention of each of the projects included in this study is to model exact postures and retrieve biomechanical information related to selected tasks. This analysis of JACK and 3DSSPP evaluates the two software products based on the researchers' data comparison, as well as the overall user experience. The researchers found that the results of the two software packages can produce different results that sometimes lead to conflicting conclusions about the safety of a given task. For example, in the analysis of Subject 2 performing Task 1 (Pose 1) the strength capability calculated at the shoulder joint varies considerably between the packages; JACK indicated that only 15% of the population will be able to perform this pose. On the other hand, 3DSSPP predicted that 87% of population is capable of performing this pose. In this case, it is unclear as to whether this posture exceeds the NIOSH strength upper limit. It is noted that some of the variability of the results may be due to the input angles and posture manipulation controlled by the user. Differences in results may be a factor of the higher degree of freedom of joint movement that is possible by manually manipulating of the postures in 3DSSPP, which may allow the manikin to pose in a sometimes unnatural and improbable manner. JACK seems to do a better job of only allowing "realistic" human contortions. In terms of biomechanics, where force calculations are critical, these differences can present conflicting results, as observed in the results of the simulations used in this study. Despite the biomechanical conflicts, the two software packages did produce relatively similar results in the ergonomic assessment of risk associated with each task. Evaluating the software's assessment of tasks, based on the overall risk score, the researchers find that the packages are relatively consistent. For the lower back, the compression forces are used to evaluate the task as falling below the NIOSH Action Limit (AL) of 3400N (denoted as green), between the AL and Maximum Permissible Limit of 6400 (denoted as yellow), or above the Maximum Permissible Limit (MPL) of 6400 (denoted as red). For the strength limits, the green zone is above the AL (more than 99% of healthy working population can perform the task), the yellow zone is between the AL and MPL (99% to 25% of a healthy working population can perform the task), and the red zone is above the MPL (less than 25% of a healthy working population can perform the task). Note that when considering the "overall" score for the strength prediction, the lowest ranking joint is used as the determinant for that entire posture.

A summary of the overall ergonomic analysis from JACK and 3DSSPP can be found in table 29.

		L4/L5 Compression Limit		Overall Strength Prediction	
		JACK	3DSSPP	JACK	3DSSPP
Subject 1	Task 1 (Pose 1)				
	Task 1 (Pose 2)				
	Task 2				
Subject 2	Task 1				
	Task 2				
	Task 3				
Subject 3	Task 1				
	Task 2				
	Task 3				
	Task 4				

Table 29. Software Comparison of Overall Ergonomic Assessment of all Tasks.

Of the 13 tasks evaluated, there was only one conflict (8%) in the overall ergonomic lower back analysis. The comparison of JACK and 3DSSPP strength capability found conflict in 3 of the 13 (23%) overall scores.

Upon evaluation of the two software packages, the researchers have made the following observations regarding the strengths and weaknesses of each product. The analysis of the software is limited to the application of the projects reported in this study, as it applies to evaluating the specific tasks aforementioned.

6.1 Learning curve

JACK requires more hours dedicated to getting acquainted with its' functions and features. Completion of several tutorials was necessary to even begin using the program. The User Manual is extensive (296 pages), compared to the 3DSSPP manual. The advantage of having such a large manual is that there are several tutorials to demonstrate the use of various aspects of the program. Conversely, it is difficult to use the program without knowledge of much of the content in the manual. The index which accompanies the JACK software is not comprehensive and the users found it difficult to find solutions for encountered errors. For 3DSSPP, a review of a basic tutorial and functions from the user manual is all that is needed to begin using the software. The User Manual is more concise (122 pages) and gives explanations of analysis reports. Its index is comprehensive and one can easily locate literature on specific subjects or problem areas.

6.2 Anthropometrics

By Default, JACK uses ANSUR (Army Anthropometric Survey) data to scale the human models. JACK also comes standard with other databases, and it has an option that allows for user definition of individual segment lengths and weights. 3DSSPP uses National Health and Nutrition Examination Survey (NHANES) data to configure the anthropometric parameters of the human model. It is possible that some variation in the moment calculations may be due to the different data used to scale the human figure. Differences in how the two software packages defined the X, Y, and Z coordinates and the positive and negative directions of the planes most certainly also affected the moment calculations.

6.3 Posture manipulation and angle input

In JACK, Human posturing techniques allow the user to quickly posture the human model while making predictions of the next movements, based on research of actual human movements and mechanics. This helped to avoid placing the manikin in an impossible posture. Joint angle manipulation was enabled for some joints (i.e., the knee), allowing the user to further manipulate the posture. Although JACK had a more intuitive inter face for posture manipulation, it was difficult to develop a system of taking angle measurements and entering them directly into the model. This meant that the majority of the posturing was created by visually referencing the photo.

3DSSPP manikin posture can be manipulated by obtaining actual angle measurements. However these measurements can be difficult to measure, unless the user has front, side and top view photos of the single pose. Another method involves using a goniometer to obtain measurements from live subjects. However, this is difficult to do while the task is being performed, as it was attempted with a live subject in the lab for the Solid Waste Collection Project. Most goniometer measurements place one arm of the goniometer along one limb, with the center at the joint and the other arm of the goniometer along the adjacent limb. This allows measurement of the angles between the two limbs. In 3DSSPP the angles requested for input are the angles taken at the joint between the limb and the horizontal. It is difficult to be sure when one is measuring a live subject that one of the arms is exactly at a 180 degree horizontal, often creating discrepancies in the input angles. In addition, it is also difficult to know if the user is recording the correct measurement when the subject is displaying asymmetric postures and poses that require twisting and lateral bending. Thus, even though input of angles is available, some of the limb orientations and postures of the virtual figure still have to be manually manipulated to attain the desired pose.

6.4 Animation

Animating the manikin in JACK with the Task Simulation Builder (TSB) is a helpful feature. This module allows the program to generate multi-step processes in a single command. For instance the "Get" command can make the human model walk over to an object, reach, and pick it up, in a single command. This feature is useful for evaluating manufacturing processes and multi-step tasks. In addition to the human animation, 3D objects can be incorporated into the scene to represent components of the environment, object being handled (i.e. materials, tools), and workstations/machinery. These objects can also be animated. Postures or tasks are input at specified time intervals or frames and the program predicts the iterations of poses and actions required to perform the task. One issue encountered with the TSB is that the software does not allow a human model to be the load, as in a victim being lifted (disaster research). However, the program does allow for multiple humans and objects to be involved in the animation. Another weakness encountered was a shift in the frame of reference for the object being lifted and inexplicable changes to the position of the hands grasping the object. This hand shifting error occurred when a manikin was selected which differed from the manikin representing the 50th percentile human. Another frequent error occurred in which the object seemed to float at odd angles through the simulation, even if the virtual human was grasping it in a pre-defined spot. The manual did not offer insight into how this problem could be corrected. It obviously was related to the matrix used as the frame of reference and the center of mass for the object; however, multiple iterations of various simulations encountered this problem.

3DSSPP also allows for animation; however, it is much more limited. The user can input postures at specified frames and the software will do some limited prediction of movement from one posture to the next. The software has some simple objects such as a box that can be scaled and placed only in the manikin's hand. 3DSSPP does not allow for multiple people to perform the tasks simultaneously, as is often the case in real-time occupational settings. Also, the "dynamic" simulation feature essentially just compiles a series data for each pose as a static load.

Overall, the information and analyses provided by JACK and 3DSSPP can be used to aid in evaluating physically intensive tasks, redesigning a task, designing products, and evaluating the ergonomic impact on a worker or user. For general occupational simulations, such as workstation design and workload distribution, exact angles aren't needed for individual poses. Simple lifting calculations can provide extremely useful ergonomic design and consideration for most occupational applications. However, where exact posture replication is desired, the user may have to employ supplemental or alternative technologies such as 3D Motion Capture.

6.5 Limitations

In the attempt to replicate unique and awkward postures, it is apparent that obtaining and applying the joint angles is an important factor in the output of the forces, moments, and strength prediction. When manipulating the manikin in 3DSSPP, the researchers noticed how a single degree in rotation of the wrist or shoulder can render the posture safe or unsafe. This fact supports the hypothesis that the variability of results between the forces and moments output may very well be due to inaccurate replication of the posture and angles, even if the postures appeared to be the same visually. Another limitation of the study is that only one camera was available, and the subject could not be simultaneously recorded from the front, side, and top view. If images from multiple planes of the same posture were obtained, then segment angles would be easier to find and replicate. This would have allowed for more accurate angle measurements. The JACK software is enabled with a Motion Capture Module that allows direct input of 3D motion data of an actual human subject. This requires expensive hardware but can give the body segment angles that are hard to manually measure. This may help to correct the human error in manually entering the angle measurements and arbitrarily manipulating the manikin's posture. One other major limitation of this study is that static postures were evaluated and not the dynamic movements of the subject. The biomechanical load on the joints may be underestimated by this limitation. Jumping, rapid acceleration and deceleration of body segments, and the duration and frequency of movement may also yield different analysis of risks and would be a necessary study for a complete comparison of the software. Dynamic Biomechanical Analysis is not an available feature of either JACK or 3DSSPP. Although JACK does allow for complex animation, it does not account for the effects of acceleration and momentum. The "dynamic" reports generated by JACK are essentially the data collected at a fixed moment in time, as in 3DSSPP, the reports are basically a series of static evaluations.

"Static Strength Prediction (SPP) is most useful for analyzing tasks that involve slow movements, since the calculations assume that the effects of acceleration and momentum are negligible." (JACK Training Manual, p. 18)

This poses an issue, especially with the Interactive Gaming study. If the program could account for speed and frequency of the motions, then a more thorough biomechanical analysis of gameplay could be observed. The RULA, NIOSH, Metabolic Energy Expenditure, and Fatigue Analysis tools provided in the JACK program could prove to give better insight into the ergonomic risk of the task, but will not provide biomechanical data.

6.6 Future areas of related research

Many software packages claim to perform biomechanical analysis of user-input data. The majority of these packages are used in sports analysis. A large percentage of the software requires specialized hardware, often requiring and expensive investment. The ideal software for use in research has yet to be identified, and may not currently exist. Features of an optimal software package would allow upload of user-supplied photos and video. The ultimate usability feature would be to allow upload of footage, including from news footage, for analysis of tasks. The ability to simulate and analyze the movements of multiple subjects simultaneously would find frequent application, especially in the disaster management realm. Graphical representation of the results, similar to the 3DSSPP output is useful to quickly identify tasks which place subjects at risk. While creating simulations, usability would be enhanced if software prompted user with suggestions to correct errors. Interactive user guides which focus on common errors and steps to correct encountered errors would be of great use to researchers and facilitate simulations. Other suggestions for further research may include exploring whether the analysis module in JACK (NIOSH, RULA, etc.) can directly use the animation data to automatically calculate output, rather than have the user manually enter the frequency, cycle time, lifting height, etc. Additionally, a Usability Study comparing the Human Modeling software may be an appropriate research topic to further expand on this study. The researchers in this study found many limitations with regard to data input and errors, as previously discussed. The learning curve for both software packages is extensive. Enhancements to the training manuals and interactive features would greatly improve the usability of both software packages and allow for a comprehensive comparative evaluation.

7. References

All Sport Systems (n.d). MotionView Video Analysis Software, In: All Sport Systems, 15 February 2011, Available from: < http://www.allsportsystems.com/>

Almedghio, S. A. (2009). Wii knee revisited: meniscal injury from 10-pin bowling, In: BMJ Case Reports, 29 July 2011, Available from: <http://casereports.bmj.com/content/2009/bcr.11.2008.1189.full>

AnyBody Technology Inc. (n.d.). AnyBody Tech Modeling Systems, In: AnyBody Technology Inc., 8 August 2011, Available from: <http://www.anybodytech.com/index.php?id=26>Barron, D. A. (2008). Wii Knee. Emergency Radiology, Vol. 15, No.4, pp. (255-257)

C-Motion, Inc. (2010). C-Motion Research Biomechanics, In: C-Motion, Inc., 15 February 2011, Available from: <http://www.c-motion.com/products/visual3d.php>

Collins, M. N. (2008). Magnetic resonance imaging of acute 'wiitis' of the upper extremity. Skeletal Radiology, Vol.37, No.5, pp. (481-3)

Das, A. (April 20 2009). More Wii Warriors Are Playing Hurt, In: NY Times, 29 July 2011, Available from: <http://www.nytimes.com/2009/04/21/health/21wii.html?adxnnl=1&partner=rss&emc=rss&adx nnlx=1302683371-q/diaU3ZZNGogDlxOaFobg>

Georgia Tech Research Institute (2011). Occupational Safety and Health Program, In: Georgia Tech Research Institute, 15 June 2011, Available from: <http://www.oshainfo.gatech.edu/index.html>

Innovision Systems INC. (n.d). MaxPRO, In: Innovision Systems INC. February 27, 2011, Available from: <http://www.innovision- systems.com/>

Jacobs, A. (2008). A Rescue in China, Uncensored, In: The New York Times, 15 June 2011, Available from: <http://www.nytimes.com/2008/05/14/world/asia/14response.html>

Kano, M., Sigel, JM., & Bourque, LB. (2005). First-aid training and capabilities of the lay public: A potential alternative source of emergency medical assistance following a natural disaster. Disasters, Vol. 29, No.1, PP. (58-74)

LA County SAR (2010). Haiti Earthquake, In: LA County SAR, 15 June 2011, Available from: <http://edwardrees.wordpress.com/2010/01/21/3-days/>

Laursen, B., Schibye, B. (2002). The effect of different sources on biomechanical loading of shoulder and lumbar spine during pushing and pulling of two-wheeled containers. Applied Ergonomics, Vol. 33, No. 2, pp. (167-174).

McNeil, S. (1999). An exploration of the Opportunities, Costs, Benefits, and issues Related to Automation of Solid Waste Collection Vehicles. Environment Research & Education Foundation.

Musculographics Inc. (2011). 15 February 2011, Available from: http://www.musculographics.com/

National Climatic Data Center, "Billion Dollar U.S. Weather Disasters". http://www/ncdc.noaa.gov/oa/reports/billionz.html, retrieved December 13, 2010.

New Jersey National Guard (n.d). Hurricane Katrina, In: New Jersey National Guard, 15 June 2011, Available from: <http://www.pdcbank.stat.nj.uc/military/publication/guardlife/volume31no5>

Siemens PLM Software (2011). JACK, In: Siemens PLM Software, 15 July 2011, Available from: <http://www.plm.automation.siemens.com/en_in/Images/4917_tcm641-4952.pdf>

Siemens PLM Software (2011). Lowback Analysis Compression Tool Background, In: Siemens PLM Software, 25 July 2011, Available from: <C:\Program Files\Siemens\Jack_7.0\library\help\TAT_Low_back.htm>

Siemens PLM Software (2011). Task Analysis Toolkit (Tat) Training Manual, In: Siemens PLM Software, 25 July 2011, Available from: <http://www.plm.automation.siemens.com/en_us/products/tecnomatix/assembly_planning/jack/index.shtml>

Siemens PLM Software (2011). Static Strength Prediction Tool Background, In: Siemens PLM Software, 26 July 2011, Available from: <file:///C:/Program%20Files/Siemens/Jack_7.0/library/help/TAT_Strength.htm#top>

Stull, J. (March, 2006), Understanding PPE selection & use during disasters. Professional Safety, Vol. 51, No.3, pp. (18-49)

University of Michigan (2010). 3D Static Strength Prediction Program, In: University of Michigan, February 27, 2011, Available from: <http://www.engin.umich.edu/dept/ioe/3DSSPP/>

University of Waterloo (n.d.). Ergowatch, In: Ergonomics and Safety Consulting Services, 8 August 2011, Available from: <http://www.escs.uwaterloo.ca/brochure.pdf>

Xcitex, Inc (2009). ProAnalyst Software, In: Xcitex, Inc, 15 February 2011, Available from: <http://www.xcitex.com/html/proanalyst_applications_examples.php>

Youngstown State University, Environmental and Occupational Health and Safety (1997). Back Belts Pros and Cons, In: Youngstown State University, 15 June 2011, Available from: <http://cc.ysu.edu/eohs/bulletins/Lifting%20Belts.htm>

5

Measurement Instruments for Ergonomics Surveys – Methodological Guidelines

Marina Zambon Orpinelli Coluci
State University of Campinas (UNICAMP)
Brazil

1. Introduction

Ergonomic surveys are very important tools to evaluate and identify problems in workplaces such as industries, hospitals, and laboratories. Strategies to tackle the ergonomic issues can be proposed based upon the results of the surveys. Therefore, the surveys should be carefully prepared to obtain information in a clear and reliable way. Usually, ergonomic surveys rely upon measurement instruments (questionnaires) that are applied to workers on the workplace to collect the necessary information.

In this chapter, we present a description of methodological guidelines used to prepare a new questionnaire or to adapt an already developed one.

The first step in developing a questionnaire is to clearly define the questions (construct) you want to answer with the ergonomic survey (Snyder et al., 2007). Based upon those questions, careful searching for questionnaires that have already been used to similar cases should be done. Having found questionnaires that measure exactly what you want, further analysis should be carried out about the questionnaire language and the sample which it was applied.

With the growth of the number of questionnaires developed for a specific culture, their use in other countries, cultures, and languages has become an important tool with the cross-cultural adaptation process (Beaton et al., 2002). Minor changes in the original questionnaire can be done to better adapt it to your purposes.

So, how to decide if it is better to use an existing questionnaire or to create a new one?

There are some advantages in using existing questionnaires: time saving in developing a questionnaire based upon steps suggested in literature; possible comparisons with previous studies involving the same questionnaire; psychometrical properties analysis in different situations; and no necessity to develop the administration and analysis processes.

Sometimes it is necessary to change some specific terms in existing questionnaires to fulfill all the requirements of the intended construct. In those cases, a content validity process should be carried out to check whether the proposed changes are misunderstood (Wynd et al., 2003).

On the other hand, when no questionnaires are found to measure the intended construct, new questionnaires can be developed. In those cases, there are steps recommended by the scientific community that guide the development of the questionnaire, such as items selection, domains development, and evaluation of the psychometric properties (Lynn, 1986; Streiner & Norman, 1995; Polit & Hungler, 1995; Turner et al., 2007; Snyder et al., 2007). In general, developing a new questionnaire is a long, laborious process. Therefore, a new questionnaire should be developed only if there are no other questionnaires for the same construct.

2. The importance of a cross-cultural adaptation process of a questionnaire if the original one was developed to be used in another language/country

Having decided to use an existing questionnaire developed in another language, it is important to carry out a cross-cultural adaptation. This adaptation allows one to apply the questionnaire for a different culture and/or tongue, and to compare results among different countries.

The term cross-cultural adaptation has been used to indicate the process that takes into account the two languages (original and adapted) and the cultural adaptation during the development of a new questionnaire to be used in different context (Beaton et al., 2002).

The cross-cultural adaptation process should follow established rules because the adaptation of a questionnaire to be used in another country, culture, or tongue needs a method to keep equivalence between the original and the adapted questionnaires (Beaton et al., 2002). The questionnaire items should be well translated and be culturally adapted to keep the validity of the instrument (Beaton et al., 2000). Guillemin (1995) have pointed out that measuring in different locations in equivalent ways is prerequisite in order to compare results from different cultures.

Before proceeding with the cross-cultural adaptation process it is necessary to request authorization for the authors of the original questionnaire regarding the use and adaptation of their instrument. The process is then completed after the following steps have been fulfilled:

a. *Translation*: this is the first step where two independent translations are recommended of the original language to the target one used in the current survey. The translations should be done by bilingual translators where the mother tongue should be the target tongue. Only one of the translators should present previous experience about the theme of the survey and may be informed about the aspects to be investigated by the survey.
b. *Synthesis*: The second step is when the two translators and the principal investigator (or a third translator) analyze and compare differences between the translations in order to synthesize the results and obtain a single, definitive version of the adapted questionnaire (Beaton et al., 2000).
c. *Back-translation*: the synthesized version should be translated back to the original language by two translators that have not participated on the first step. Their mother tongue should be the one of the original questionnaire and should not be informed about concepts to be explored within the instrument. These translators do the translations independently, without previous knowledge of the original questionnaire.
d. *Content evaluation*: After the first three steps, an expert committee is organized to evaluate the content of the questionnaire. This committee is composed by bilingual

professionals with large experience on the topics covered by the questionnaire. The professionals receive the translations, the synthesis, the back-translations, and instructions about how to carry out the evaluation of the questionnaire content. After a detailed analysis, the professionals produce a pre-final version of the adapted questionnaire.

e. *Pre-test*: With the pre-final version of the questionnaire, a pre-testing is carried out in a sample of typically 40 subjects (Beaton et al., 2000). Each of the subjects fill the questionnaire and is interviewed about the understanding of the items, words, and easiness of the filling the questionnaire. During this step, the subjects can point out difficulties and suggest modifications to improve the instrument. If the suggested changes are significant and extensive, another analysis of the expert committee is necessary. At the end of this step a final version of the adapted questionnaire is obtained.

Researchers are following these steps when performing a cross-cultural adaptation process (Vigatto et al., 2007; Gallasch et al., 2007; Toledo et al., 2008; Coluci & Alexandre, 2009; Coluci et al., 2009) and it is possible to verify that they used carefully methods in order to conduct the process in a reliable way.

It is important to note that, often, one can find in the literature questionnaires that measure the construct to be evaluated with good psychometric properties. After permission of the original authors of the questionnaire, it is possible to use it without making modifications if the recommendations presented in the instrument are followed.

However, one must be careful when using an instrument ever built. When it was created, were the psychometric properties evaluated with the same population you intend to study?

If the answer to this question is "yes", you can use the questionnaire with greater tranquility, but you must verify whether the cultural context and the situation are similar to yours.

If the answer is "no", you should evaluate the psychometric properties of this questionnaire to the other population. This probably can occur when you choose to use a questionnaire to assess a construct in a generic form, i.e., when it is not designed to a specific population. An example of this situation is the study conducted by Shimabukuro et al. (2011), which aimed to adapt a generic questionnaire that evaluates the workers' perception regarding job factors that can contribute to musculoskeletal symptoms to physical therapists. The authors made some changes in the questionnaire's content and evaluated the psychometric properties with the specific population.

And why is it important to check these properties again? It is simple. Applying a questionnaire to a population different from that involved in the study during its development process, one can find different results (better or worse) than the original. Therefore, such assessment can demonstrate if the questionnaire is also reliable and valid for the other population.

3. A description of all steps for developing a new measuring instrument

When a new questionnaire is necessary, researchers should follow standard and systematic methods that aim to improve the quality of measuring instruments (Haynes et al., 1995; Keszei et al., 2010; Pittman & Bakas, 2010).

The following steps are suggested: definition of the conceptual structure; definition of the target population and the objectives of the instrument; development of the domains and selection of the items; organization of the instrument; evaluation of the content validity and pre-test; and finally the evaluation of the psychometric properties.

a. *Definition of the conceptual structure*: This step aims to help an initial development of the items and domains. Some methods can be used in this stage such as literature search, interviews with specialists in the field and/or with subjects of the target population, focus groups, other questionnaires analysis, and meetings with a referee committe (Benson & Clark, 1982; Berk, 1990; Turner et al., 2007).
b. *Definition of the target population and the objectives of the instrument*: It is important to characterize the target population in order to justify the relevance of a specific questionnaire (Turner et al., 2007). It is also fundamental to establish a link between the concepts involved and the development of the questionnaire (Fagarasanu & Kumar, 2002).
c. *Development of the domains and selection of the items*: The domains to be investigated with the questionnaire are listed based on the relevance of the proposed survey (Snyder et al., 2007). The selection of the items of the questionnaire can be obtained through literature search and interviews with subjects of the target population and specialists in the field (Streiner & Norman, 2002; Turner et al., 2007). The literature search should be carried out in databases, looking for related constructs and questionnaires in order to determine reference constructs. The interviews with the target population aim to determine individual perceptions about the involved aspects and provide important preliminary data during the development of the questionnaire. The interviews with specialists allow to verify the content to be explored with the questionnaire.
d. *Organization of the instrument*: At this step, the items are organized in their respective domains and a final form for the questionnaire is prepared which includes title, instructions, and response scale. The response scale type and scores are determined based upon the easiness for understanding and answering by the subjects, and evaluating by the researchers (Turner et al., 2007).
e. *Evaluation of the content validity*: This is an essential step in the development of a new questionnaire. It allows associating abstract concepts with measurable and observable quantities (Kirshner & Guyatt, 1985). Details of this step will be provided in section 8 of this chapter.
f. *Pre-test*: The pretest should be applied in a sample of the population in order to verify the understanding of the new questionnaire. After the administration of the questionnaire, the investigator should interview each subject individually and ask him/her about the understanding of words and items as well as about the procedures of filling in their answers. Modifications can be made according to the suggestions of these subjects. When the changes are significant, it is important to be evaluated and approved again by the expert committee that carried out the content validity. After this phase, the measuring instrument is completed and its psychometric properties can be studied.
g. *Evaluation of the psychometric properties*: The evaluation of the psychometric properties of a new questionnaire is one of the most important steps because it allows verifying the validity and reliability of the instrument to be used in other research and/or ergonomics practices. When we create a questionnaire, we intend to disclose it to the scientific community. If the questionnaire shows good psychometric properties, it can

be widely used by other researchers. Therefore, its use can be widespread whether it is well constructed and evaluated.

The techniques to verify the psychometric properties will be explained in sections 8 and 9 of this chapter.

It can be noted that recent studies involving the development of new questionnaires are following these steps (Farias et al., 2008; Buysse et al., 2010; Bergman et al., 2011; Giesler et al., 2011; Marant et al., 2011; Young et al., 2011). These studies showed the steps of literature review on the topic being discussed and literature review on other scales that could be used for the same purpose. Furthermore, some researchers consulted experts with experience on the area of interest during the selection of domains and items (Farias et al., 2008; Bergman et al., 2011); others conducted focus groups and semi-structured interviews to obtain relevant information for the generation of items (Buysse et al., 2010; Young et al., 2011); and others have conducted interviews with a sample of the target population in order to obtain important suggestions during the developing of the conceptual model of the questionnaire (Giesler et al., 2011; Marant et al., 2011).

4. Content validity – How to do and how to evaluate this validity using qualitative and quantitative methods?

There are controversies about the terminology and the concept of content validity (Sireci, 1998; Haynes et al., 1995). For some authors, content validity is associated in determining in which fraction the selected items represent appropriately the important aspects of the concept to be evaluated (Contandriopoulos et al., 1999). It aims to verify the extension of the items that determine the same content (Rubio et al., 2003). For other authors, the content validity is an answer for the following question: Are the items of the questionnaire representative among all the questions that can be formulated about the topic in analysis? (Polit & Hungler, 1995).

Another way to define the content validity is the process to evaluate the degree of relevance and representativeness of each element of the questionnaire with respect to a specific construct (Haynes et al., 1995). The elements of the questionnaire include the items, instructions, and format of the answers because all of them can influence the data collection.

For some authors the content validity comprises only the evaluation by an expert committee (Dempsey & Dempsey, 1996; Fitzner, 2007). However, the content validity has been described as judgment process composed by two distinct parts: (i) the development of the questionnaire, and (ii) its evaluation by an expert committee (Lynn, 1986; Polit & Beck, 2006).

The number and qualification of the judges of the committee is controversial. Lynn (1986) suggests a number between 5 and 10 whereas Haynes et al. (1995) suggest a number between 6 and 20 with groups of at least 3 individuals of each field. Other aspects such as the characteristics of the questionnaire, formation, qualification and availability of the judges can be taken into account (Lynn, 1986; Grant & Davis, 1997).

Different criteria can be used to select the group of specialists such as clinical experience, research and publication on the field, expertise on the involved conceptual structure, and methodological knowledge about development of questionnaires and scales (Berk, 1990;

Grant & Davis, 1997). It is also suggested the participation of lay persons related to the target population of the questionnaire (Tilden et al., 1990; Rubio et al., 2003).

In cases involving cross-cultural adaptation, a multidisciplinary committee is suggested (Hutchinson et al., 1996). In this case, the committee would be formed by bilingual specialists that know the concepts and measures involved (Guillemin et al., 1993).

The evaluation by the judges can involve both quantitative and qualitative procedures (Tilden et al., 1990; Burns & Grove, 1997; Hyrkäs et al., 2003). The process begins with an invitation of the judges that receive instructions and a specific questionnaire for the evaluation (Grant & Davis, 1997). A letter of invitation should explain the reason because the specialist was chosen, the relevance of the involved concepts, and overall explanation of the questionnaire (Lynn, 1986; Grant & Davis, 1997), including the aim of the survey, the scales used, and the adopted score (Davis, 1992; Rubio et al., 2003). The letter can also include conceptual and theoretical foundations from the questionnaire (Davis, 1992) and information about the target population. If lay persons will compose the committee, a description of the educational level of the members can be specified in the letter (Rubio et al., 2003).

Initially, the judges should analyze the coverage of the questionnaire, i.e., if each domain has been covered by the selected set of items (Tilden et al., 1990). In this stage, the committee can include or remove items of the questionnaire (Rubio et al., 2003). Then, a detailed analysis of the items is performed individually. The committee should evaluate the clarity on the writing of each item to guarantee that each item is not misunderstood (Grant & Davis, 1997). The committee also should analyze if the number of items are adequate and relevant to reach the aims of the survey (Grant & Davis, 1997; McGilton, 2003). Suggestions by the judges to improve specific items can be done at this stage (Tilden et al., 1990; Rubio et al., 2003).

The dynamics of the evaluation process by the judges can occur either individually by each judge followed by a group discussion or interactively through interviews and discussions about the controversial points (Grant & Davis, 1997).

To quantify the level of agreement among the specialists during the evaluation of the content validity, different methods can be used:

a. *Percent agreement score*: The agreement between the specialists (in percentage) is quantified by the ratio of the number of specialists that agree with each other and the total number of specialists (Tilden et al., 1990; Hulley et al., 2003). This is the simplest method to determine the level of agreement (Topf, 1986) and has been used on the initial determination of the items (Tilden et al., 1990; Grant & Davis, 1997). The simplicity in the calculation is an advantage of this method. However, some limitations forbid the use of this method in all cases (Topf, 1986). This methods should be used considering an agreement of 90% among the specialists (Topf, 1986; Polit & Beck, 2006).
b. *Content validity index*: This method quantifies the proportion of judges that agree about some specific aspect of the questionnaire and its items. It is used commonly on the health field (Wynd & Schaefer, 2002; Hyrkäs et al., 2003; McGilton, 2003).

The method allows analyzing each item individually and also the questionnaire as a whole through the use of a Likert-like scale with score from 1 to 4. The numbers express the level of changes/understanding the judge had about the item. For example, the following definitions can be applied: (i) 1 = not representative, 2 = needs major revision to become representative,

3 = needs minor revision to become representative, 4 = representative (Lynn, 1986; Rubio et al., 2003), or (ii) 1 = not clear, 2 = unclear without item revision, 3 = clear but needs minor modifications, 4 = very clear (Hyrkäs et al., 2003; Wynd et al., 2003; DeVon et al., 2007).

The content validity index for each item of the questionnaire is then calculated by the ratio of the number of answer with scores "3" and "4" and the total number of answers (Grant & Davis, 1997; Wynd et al., 2003). Items with score "1" and "2" should be revised or even removed.

To evaluate the questionnaire as a whole, different ways can be used. For instance, Polit and Beck (2006) presented three ways: (i) use of the average of the proportions of the items considered by the specialists; (ii) use of the sum of all indexes calculated separately divided by total number of items analyzed; and (iii) use of the ratio of the total number of items considered as relevant by the specialists and the total number of items.

It is also important to define acceptable agreement rate. Some authors consider the number of specialists on the evaluation of the individual items. When the number of specialists is less than 5, all should agree (rate equal to 1) for an item to be considered as relevant. For a number of 6 or more specialists, the rate should not be less than 0.78 (Lynn, 1986; Polit & Beck, 2006). Some authors suggest a minimal rate of 0.80 to check the validity of new instruments (Davis, 1992; Grant & Davis, 1997) however the recommended rates should be larger than 0.90 (Polit & Beck, 2006).

c. *Kappa coefficient*: The kappa coefficient is the ratio of the proportion of the number of specialists that agreed and the maximum proportion that the specialist could agree (Hulley et al., 2003, Siegel & Castellan, 2006). It is useful when the data are divided in categories and represented nominally (Siegel & Castellan, 2006). The values of kappa are in the range of -1 (no agreement) to 1 (total agreement) (Hulley et al., 2003).

5. The reliability assessment: Importance and procedures to evaluate it in a new questionnaire

Reliability is the ability to consistently reproduce a result in time and space, or using different observers (Contandriopoulos, 1999). It indicates aspects about the questionnaire coherence, precision, stability, equivalence, and homogeneity (Lobiondo & Haber, 2001).

It can be evaluated by three different methods: the stability (test-retest), the homogeneity, and the equivalence (inter-observer).

The stability aims to analyze the consistency of the instrument when repeating the measures using a test-retest design (Polit & Hungler, 1995). When you decide to use this method, the situation that is being measured must be in the same conditions in both test, and some differences between tests must be due to random errors (Burns & Grove, 1997).

The homogeneity or internal consistency can be evaluated to verify whether all items of a questionnaire are related to different aspects of the same construct (Streiner & Norman, 1995). Using this method, you can verify if the questions of the instrument measure the same concept (Lobiondo & Haber, 2001).

The equivalence reliability is an inter-observer measure and it allows verifying whether the administration of a specific instrument by two different persons will provide the same results.

6. Validity concepts: Types, importance, and procedures to evaluate the psychometric properties

Validity is an important psychometric property used to evaluate the quality of an instrument (Polit & Hungler, 1995). It is related to the fact that a questionnaire should really measure what is intended to, i.e., the validity can show if the questionnaire represents the concept that it is trying to measure (Lobiondo & Haber, 2001).

Content validity was already defined in section 8 and it aims to analyze whether the questionnaire items are relevant to measure the proposed content.

Criterion validity is used when there is a "gold standard" questionnaire to compare with your questionnaire. This method indicates whether the results obtained with the target questionnaire corresponds to the results obtained with another observation/instrument that measures the same content of interest (Guillemin, 1995).

Construct validity is one of the most important characteristics of an instrument because it evaluates how much the instrument measures the construct of interest. It involves the generation of a hypothetical model to describe the constructs to be assessed and to determine their relationships (Fayers & Machin, 2000).

This type of validity covers a variety of techniques aimed, therefore, to assess whether the theoretical construct appears to be an appropriate model and whether the measuring instrument corresponds to the construct.

Factor analysis is one of the most important and powerful methods to establish the construct validity (Fayers & Machin, 2000). This type of analysis allows us to establish whether there is strong correlation between variables within the same group, but weak correlations between variables from outside the group (Fayers & Machin, 2000).

The factor analysis can be exploratory when you are developing a new questionnaire and there is no prior knowledge of the structure to be used, i.e., it creates a structure for the instrument (Fayers & Machin, 2000). It can also be confirmatory when the goal is to test whether the correlations correspond to the predefined structure of the questionnaire, confirming the number of items previously developed as well grouping the items into factors or domains (Fayers & Machin, 2000).

There is also the construct validity that uses the known-group technique, which consists of looking for different results when applying a questionnaire to groups with contrasting characteristics (Polit & Hungler, 1995; Dempsey & Dempsey, 2000).

The convergent validity is also another technique to verify the construct validity. It consists in showing that a dimension of the new instrument correlates with other dimensions of questionnaires theoretically related (Fayers & Machin, 2000). In contrast, the divergent validity assesses the questionnaire domains correlating them to other domains of questionnaires which content should not be related to the investigation.

Depending upon the type of the questionnaire, we should choose different techniques to evaluate the reliability and validity. This choice should be based on the availability of key technical aspects for each type of technique. For example, if you are developing a new measuring instrument which construct was not measured by any other questionnaire, probably you will not be possible to perform criterion validity.

7. The questionnaire application methods and the procedures to decide the better way to assess a population

During the development of a new questionnaire, the researcher has to think about how to apply it. The type of application method can influence which questions can be asked and in what format (Streiner & Norman, 2002). It is possible to choose one of these four types of methods: face-to-face interviews, self-administration, over the telephone, and by mail.

a. *Face-to-face interviews:* This method is used when the author decides to interview each subject individually. The researcher must recruit each subject, explain the importance of research and how to proceed, and, from the consent of the subject, perform the questions and record the answers of the subjects. This method has the advantage of a greater participation of the subjects because the researcher has the opportunity to personally explain the importance of his/her study. In addition, the researcher can clarify doubts during the administration of the questionnaire when the subject demonstrates any difficult on answering it.

It is important to consider another aspect of this method. If the researcher has any link with the research site or any of the subjects who participate in the study, it is recommended that the researcher do not conduct the interviews. In order to minimize any interference in the responses of the subjects, the researcher must instruct another person to apply the survey. This person must be able to answer any questions presented by the subjects.

b. *Self-administration:* This technique can be chosen when the researcher has sufficient knowledge whether the subjects are able to answer the questionnaire by themselves. Therefore, one should consider the educational level of the population studied and whether the terms used in the questionnaire will be understood by the subjects. You can apply this technique in two ways. In both the researcher can explain the importance of the survey in person and give instructions on how to complete the survey. Then, the researcher can choose to leave the questionnaire with the subject and set a date and time to collect it. Or the researcher can ask the subject to answer the questionnaire in his/her presence. The disadvantage of this method is that the researcher can not clarify any doubt of the subjects, even if the researcher is present. The advantage is that there is less bias to answer, i.e., less interference from the researcher in the subject's response.
c. *Over the telephone:* This method can be an interesting alternative when there is difficulty in performing a presence interview. There are some advantages such as reduction of blank answers, clarification of doubts, and recruitment of a larger number of subjects for participation in the research. However, there may be difficulty in obtaining the informed consent of subjects for study participation and some people may suspect the intention of the researcher as they do not see him/her personally. In addition, the questionnaire applied over the telephone can be useful when the questionnaire has only open-ended questions and when it is not too long, as most people do not appreciate to stay long time on the telephone.
d. *By mail:* This technique can be the cheapest one and it allows the recruitment of a large number of subjects for participation in research. You can also send along with the questionnaire a formal request for written consent of the subjects. In addition to these documents, a letter explaining the importance of research and how the subject should respond to the questionnaire should also be included. However, the most important disadvantage of this method is the highest number of denied participation in the

research. It is almost impossible to recover the questionnaire whether the subject, even with reminders sent by the researcher, does not return the instrument. Another disadvantage is the number of blank or invalid answers, because the subject can try to answer the questionnaire in a sequence different from what the researcher would like and this may influence the responses.

8. Summary

This chapter provides useful information for researchers interested in evaluating surveys on ergonomics. The instrument used to obtain data - questionnaires - should be carefully chosen based on the target population, constructs intended to be measured, existence of similar questionnaires, methods of administration, and psychometric properties. For questionnaires previously developed for a different language and/or culture, the chapter also presents the steps to a cross-cultural adaptation. If a new questionnaire is really necessary, which is decided after a careful analysis, the procedures to develop it are also explained. Finally, in order to show that the questionnaire is suitable for the target population and whether it measures what is intended to, the types of evaluation of the psychometric properties - reliability and validity - are described.

9. References

Beaton DE, Bombardier C, Guillemin F & Ferraz MB. (2000). Guidelines for the process of cross-cultural adaptation of self-report measures. *Spine,* Vol. 25, No. 24, pp. (3185-91).

Beaton D, Bombardier C, Guillemin F & Ferraz MB. (2002). Recommendations for the cross-cultural adaptation of health status measures. *Am Acad Orthop Surg,* pp. (1-9).

Benson J & Clark F. (1982). A guide for instrument development and validation. *Am J Occup Ther,* Vol. 36, No. 12, pp. (789-800).

Bergman HE, Reeve BB, Moser RP, Scholl S & Klein WMP. (2011). Development of a comprehensive heart disease knowledge questionnaire. *Am J Health Educ,* Vol. 42, No. 2, pp. (74–87).

Berk RA. (1990). Importance of expert judgment in content-related validity evidence. *West J Nurs Res,* Vol. 12, No. 5, pp. (659-71).

Burns N & Grove SK. (1997). *The practice of nursing research.* (3rd ed.), Saunders, Philadelphia.

Buysse DJ, Yu L, Moul DE, Germain A, Stover A, Dodds NE & et al. (2010). Development and validation of patient-reported outcome measures for sleep disturbance and sleep-related impairments. *SLEEP,* Vol. 33, No. 6, pp. (781-92).

Coluci MZO & Alexandre NMC. (2009). Cross-cultural adaptation of an instrument to measure work-related activities that may contribute to osteomuscular symptoms. *Acta Paul Enferm,* Vol. 22, No. 2, pp. (149-54).

Coluci MZO, Alexandre NMC & Rosecrance J. (2009). Reliability and validity of an ergonomics-related Job Factors Questionnaire. *Int J Ind Ergon,* Vol. 39, No. 6, pp. (995-1001).

Contandriopoulos AP. (1999). *How to prepare a research.* (3rd ed.), Hucitec/Abrasco, São Paulo.

Davis LL. (1992). Instrument review: getting the most from a panel of experts. *Appl Burs Res,* Vol. 5, pp. (194-7).

Dempsey PA & Dempsey AD. (1996). *Using nursing research.* (5th ed.), Lippincott, Philadelphia.

Dempsey PA & Dempsey AD. (2000). *Using nursing research: process, critical evaluation, and utilization*. (3 ed.), Lippincott, Philadelphia.

DeVon HA, Block ME, Moyle-Wright P, Ernst DM, Hayden SJ, Lazzara DJ & et al. (2007). A psychometric toolbox for testing validity and reliability. *J Nurs Scholarsh*, Vol. 39, No. 2, pp. (155-64).

Fagarasanu M & Kumar S. (2002). Measurement instruments and data collection: a consideration of constructs and biases in ergonomics research. *Int J Ind Ergon*, Vol. 30, pp. (355-69).

Farias ST, Mungas D, Reed BR, Cahn-Weiner D, Jagust W, Baynes K & et al. (2008). The measurement of everyday cognition (ecog): scale development and psychometric properties. *Neuropsychology*, Vol. 22, No. 4, pp. (531–44).

Fayers PM & Machin D. (2000). *Quality of life: assessment, analysis and interpretation*. John Wiley & Sons Ltd, Chichester.

Fitzner K. (2007). Reliability and validity. *Diabetes Educ*, Vol. 33, No. 5, pp. (775-80).

Gallasch CH, Alexandre NMC & Amick B. (2007). Cross-cultural adaptation, reliability, and validity of the Work Role Functioning Questionnaire to Brazilian Portuguese. *J Occup Rehabil*, Vol. 17, No. 4, pp. (701-11).

Giesler M, Forster J, Biller S & Fabry G. (2011). Development of a questionnaire to assess medical competencies: reliability and validity of the questionnaire. *GMS Zeitschrift für Medizinische Ausbildung*, Vol. 28, No. 2, pp. (1-15).

Grant JS & Davis LL. (1997). Selection and use of content experts for instrument development. *Res Nurs Health*, Vol. 20, pp. (269-74).

Guillemin F, Bombardier C & Beaton D. (1993). Cross-cultural adaptation of health-related quality of life measures: literature review and proposed guidelines. *J Clin Edipemiol*, Vol. 46, No. 12, pp. (1417-32).

Guillemin F. (1995). Cross-cultural adaptation and validation of health status measures. *J Rheumatol*, Vol. 24, pp. (61-3).

Haynes SN, Richard DCS & Kubany ES. (1995). Content validity in psychological assessment: a functional approach to concepts and methods. *Psychol Assess*, Vol. 7, No. 3, pp. (238-47).

Hulley SB, Cummings SR, Browner WS, Grady D, Hearst N & Newman TB. (2003). *Outlining the clinical research*. (2nd ed.), Artmed, Porto Alegre.

Hutchinson A, Bentzen N & König-Zahn C. (1996). *Cross cultural health outcome assessment; a user´s guide*. ERGHO, The Netherlands.

Hyrkas K, Appelqvist-Schmidlechner K & Oksa L. (2003). Validating an instrument for clinical supervision using an expert panel. *Int J Nurs Stud*, Vol. 40, pp. (619-25).

Keszei A, Novak M & Streiner DL. (2010). Introduction to health measurement scales. *J Psychosom Res*, Vol. 68, No. 4, pp. (319-23).

Kirshner B & Guyatt G. (1985). A methodological framework for assessing health indices. *J Chron Dis*, Vol. 38, No. 1, pp. (27-36).

Lobiondo G & Haber J. (2001). *Research in nursing: methods, critical evaluation, and utilization*. (4th ed.), Guanabara Koogan, Rio de Janeiro.

Lynn MR. (1986). Determination and quantification of content validity. *Nurs Res*, Vol. 35, No. 6, pp. (382-5).

Marant C, Arnould B, Marrel A, Spizak C, Colombel JF, Faure P & et al. (2011). Assessing patients' satisfaction with anti-TNFα treatment in Crohn's disease: qualitative steps of the development of a new questionnaire. *Clin Exp Gastroenterol*, Vol. 4, pp. (173–80).

McGilton K. (2003). Development and psychometric evaluation of supportive leadership scales. *Can J Nurs Res*, Vol. 35, No. 4, pp. (72-86).

Pittman J & Bakas T. (2010). Measurement and instrument design. *J Wound Ostomy Continence Nurs*, Vol. 37, No. 6, pp. (603-7).

Polit DF & Hungler BP. (1995). *Fundamentals of nursing research.* (3rd ed.), Artes Médicas, Porto Alegre.

Polit DF & Beck CT. (2006). The content validity index: are you sure you know what's being reported? Critique and recomendationas. *Res Nurs Health*, Vol. 29, pp. (489-97).

Rubio DM, Berg-Weger M, Tebb SS, Lee S & Rauch S. (2003). Objectifying content validity: conducting a content validity study in social work research. *Soc Work Res*, Vol. 27, No. 2, pp. (94-105).

Shimabukuro VGP, Alexandre NMC, Coluci MZO, Rosecrance JC & Gallani MCJB. (2011). Validity and Reliability of a "Job Factors Questionnaire" related to the working tasks of physical therapists. *Int J Occup Saf Ergon* (in press).

Siegel S & Castellan HJ. (2006). *Nonparametric statistics for the behavioral sciences.* (2nd ed), Artmed, Porto Alegre.

Sireci SG. (1998). The construct of content validity. *Soc Indic Res*, Vol. 45, pp. (83-117).

Snyder CF, Watson ME, Jackson JD, Cella D, Halyard MY & Mayo/FDA Patient-Reported Outcomes Consensus Meeting Group. (2007). Patient-reported outcome instrument selection: designing a measurement strategy. *Value Health,* Vol. 10(Suppl. 2), pp. (S76–85).

Streiner DL & Norman GR. (1995). *Health measurement scales: a practical guide to their development and use.* (2ed.), Oxford University Press, New York.

Tilden VP, Nelson CA & May BA. (1990). Use of qualitative methods to enhance content validity. *Nurs Res*, Vol. 39, No. 3, pp. (172-5).

Toledo RCMR, Alexandre NMC & Rodrigues RCM. (2008). Avaliação das qualidades psicométricas de uma versão brasileira do Spitzer Quality of Life Index em pacientes com dor lombar. *Rev Lat Am Enfermagem*, Vol. 16, No. 6, pp. (943-50).

Topf M. (1986). Three estimates of interrater reliability for nominal data. *Nurs Res*, Vol.35, No. 4, pp. (253-5).

Turner R, Quittner AL, Parasuraman BM, Kallich JD, Cleeland CS & Mayo/FDA Patient-Reported Outcomes Consensus Meeting Group. (2007). Patient-reported outcomes: instrument development and selection issues. *Value Health,* Vol. 10(Suppl. 2), pp. (S86–93).

Vigatto R, Alexandre NMC & Correa Filho HR. (2007). Development of a Brazilian Portuguese version of the Oswestry Disability Index: cross-cultural adaptation, reliability, and validity. *Spine*, Vol. 32, No. 4, pp. (481-6).

Wynd CA & Schaefer MA. (2002). The Osteoporosis Risk Assessment Tool: establishing content validity through a panel of experts. *Appl Nurs Res*, Vol. 16, No. 2, pp. (184-8).

Wynd CA, Schmidt B & Schaefer MA. (2003). Two quantitative approaches for estimating content validity. *West J Nurs Res*, Vol. 25, No. 5, pp. (508-18).

Young JM, Walsh J, Butow PN, Solomon MJ & Shaw J. (2011). Measuring cancer care coordination: development and validation of a questionnaire for patients. *BMC Cancer*, Vol. 11, pp. (298).

6

Biomechanical Assessment of Lower Limbs Using Support Moment Measure at Walking Worker Assembly Lines

Atiya Al-Zuheri, Lee Luong and Ke Xing
University of South Australia, School of Advanced Manufacturing and Mechanical Engineering, Mawson Lakes, South Australia, Australia

1. Introduction

Manual assembly line work is currently still necessary in the manufacturing industry. The human body despite its organic limitations is still more flexible than machines, and the human mind possesses creative and intuitive functions above that of robotic devices. Automation and robotic cells have limitations and manual assembly lines are considered a significant and justifiable solution (Hunter, 2002). In traditional assembly lines, such as Fixed Worker Assembly Lines (FWAL), each worker has a designated task, and is required to continuously repeat that task. Although FWALs are efficient and generally reliable, they have the following deficiencies (Wang et al., 2005):

- Low flexibility (in terms of workers and products),
- Need constant attention and management, and
- Difficult balancing.

It is essential that assembly systems are flexible, in order to respond adequately to the changeable characteristics and demands of the market. These demands are typically; an increasing customisation of product, shortening of a product lifecycle, and highly varied production of small batches of product (Miyake, 2006). For this reason, it has become necessary to develop dynamic, flexible and reconfigurable assembly systems. The flexible manpower line (or flexible assembly line), is one of the promising techniques for configuring effective and productive assembly systems, responding well to the challenges of the manufacturing industry (Stockton et al., 2005). It focuses on work force as key resources due to their flexibility and creativity. An example of such systems is so-called Walking Worker Assembly Line (WWAL), in which each worker utilizes various skills and functions by travelling along the manufacturing line to carry out all the required tasks.

2. Description of the WWAL

In last 15 years, several researchers have treated the topic of multifunctional walking (moving) workers performance, in production systems. Wang, Owen and Mileham (Wang et al., 2005) and Nakade and Nishiwaki (Nakade & Nishiwaki, 2008) gave a summary of this research. In all of this research, application of moving multifunctional workers was found to be limited to a cell in linear or U-shaped production lines. In addition, most of this research referred to the systems under scrutiny by various different names than WWAL.

The term WWAL is recent concept (Wang et al., 2005; Bley et al., 2007). The term is usually used to designate workstations configuration as horizontal "U" shape or straight line layouts. Each multifunctional worker travels by walking down the line carrying out each assembly task at each workstation as scheduled. Thereby, each walking worker completes the assembly of a product in its entirety from start to completion. Figure 1 illustrates concept of WWAL, where a walking worker completes a product assembly process at the last workstation $K\,\hat{}$ and then moves back to the first workstation 1 to begin the assembly of a new product.

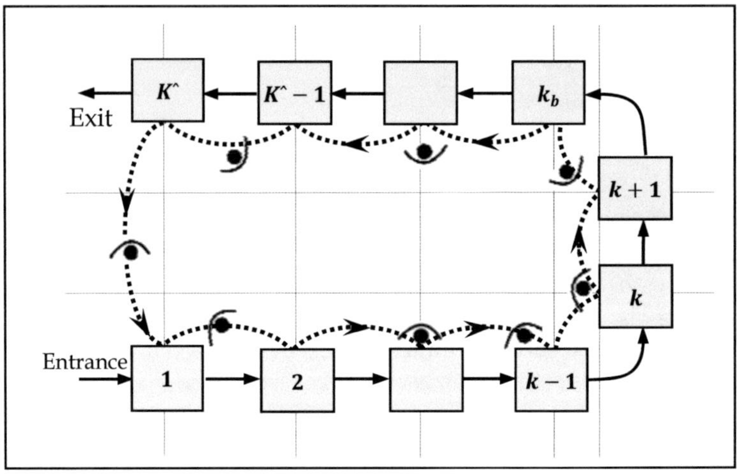

Fig. 1. Form of the walking worker assembly line.

2.1 Workstations and tools

The nature of assembly process at most of workstations in WWAL requires a manual task to be performed by the worker, using simple hand powered equipment such as trimming, riveting and fastening tools...etc. This type of workstation limits worker input to the loading and tooling of components for the end product, prior to the next step in the process. Work-pieces are loaded into a specially designed fixture. The work-pieces then put through a fixed cycle of operations using a predefined range of tools.

The process operations at each workstation are relatively small and highly specific to individual components, utilizing the specialized skills of the worker. The set-up process is relatively quick, thus losing little or no time in non-productive activity, consequently it is more efficient and cost-effective.

2.2 The workers

The workers in WWAL operate to an unaltered, repetitive sequence in which they carry out manual tasks. These tasks consist of the picking up or installing parts, or picking up and using tools and incorporates quality checks or inspections at certain stages of production.

This repetitive sequence is known as a worker operating time. The time taken to complete a worker operating time sequence is known as the overall cycle time. It is the sum of the times required to perform manual tasks, the walking times between the different workstations and in-process waiting time (if exists') of the worker at the bottleneck workstation on the line. Manual transport of components between the workstations of the assembly line requires that the time and energy required doing so, be reduced as far as possible. This is often achieved by shortening the distances between workstations. The WWAL is designed to be able to run effectively with more or fewer workers. The capacity to adjust staffing levels to suite varying required production volumes, is the key to the ability of these lines in response to changes in demand.

2.3 Line layout design

Three types of layout using multifunctional walking (moving) workers have been identified by the authors in terms of the system layout design (Wang et al., 2009):

1. U-shaped design,
2. Straight line design, and
3. L-shaped design.

The U-shaped design is perhaps the most common layout used to implement WWAL. Organizing the WWAL along U-shaped layout eliminates virtually all Work-In-Process (WIP). There are small spaces between workstations to enable a worker with a partially assembled product to queue on reaching the next workstation (if the worker arrives during the time that the preceding worker is still operating at that workstation) until it becomes empty.

Reducing space for in-process waiting enables workstations to be placed very close to one another, thereby reducing the amount of energy and time expended, increasing performance, and efficiently utilizing the available floor space. Close spacing also means products and the rest of the line are more visible to the workers, and is considered to have a beneficial effect on morale. Workers are able to see the progress of parts through the entire line, rather than at just one operation. In addition, shortened travelling distance has other inherent benefits beyond efficiency improvements; not only increased visibility of active areas, but ease of communication and increased teamwork among workers (Grassi et al., 2004; Al-Zuheri et al., 2010a).

2.4 Experiences of manufacturing companies with WWAL

Walking worker assembly line results in a series advantages over a traditional line—FWAL. In this context, rearranging assembly lines from the FWAL to the WWAL by a number of companies has led to achieve the following (Mileham et al., 2000):

1. Increased ease in line balancing, thus reducing the number of buffers required,
2. Flexible and optimal adjustment of the number of line workers to suite output demand, and
3. Minimizing the cost of labour and tooling.

Bischak (Bischak, 1996) and Zavadlav E et al. (Zavadlav et al., 1996) investigated using (moving) walking workers approach in the case of variability in operation times is high (e.g.

manual assembly line). Both found that WWAL gives the best expected production and FWAL is the worst. WWAL system and process design provides workers more control over the speed of the production process and encourages focussed attention to detail, ensuring higher work quality, and hence higher overall product quality. Deploying a WWAL approach also provides increased ergonomic benefits reducing potential muscular-skeletal problems in jobs where single and repetitive tasks are required of static workers. Increased freedom of movement, in particular walking, by the worker in WWAL systems can reduce the probability of Work-Related Muscular-Skeletal Disorders (WMSD), in the arms, back and shoulders (Moller et al., 2004).

Although implementation of WWAL systems offers a variety of benefits to manufacturers (as stated above), it has yet to be widely adopted within the industry. In this regard, Miltenburg (Miltenburg, 2001) stated that U-lines with more than one multi-skilled walking worker rarely run in chase mode, (another name of organising walking workers in this way). Only 1.3% of the U-lines deploying numerous multi-skilled walking workers use this system of production. The Japanese management institute (Gemba Research and Kaizen Institute) interpreted the lack of WWAL deployment in industrial environments to assertions by some practitioners that it has certain aspects detrimental to labour productivity and ergonomic conditions (Miller, 2007). This was mainly due to two main reasons; firstly, adopting WWAL in assembly processes, requires multifunctional workers. These have specialised skills and cost more to employ. Secondly, there is some question as to whether workers actually keep up with completing all required production steps in one cycle time. This claim is based on the time used standing and for carrying in-process products to each process point.

Undoubtedly, the question arises as to whether or not workers will have the endurance to complete a shift time of eight hours, and still have enough energy for a normal life after work. Furthermore, existing research about WWAL or similar dynamic systems (e.g. cellular system) provides only incomplete data modelling for WWAL ergonomics from which to assess the relevant concerns of practitioners about the health and wellbeing of the WWAL work force.

3. Workers postures in WWAL: Implications and investigation

3.1 Workers postures and their implications for workers

Like other manual works in industrial assembly, the tasks of WWAL include lifting, carrying, pushing, pulling of materials, and quality control. Sometimes such work requires frequently lifting heavy loads. This may include the use of non-powered or power hand tools. In addition to that, it may have long cycle and excessive walking time including load carrying (Melin et al., 1999). In general, this work involves postures that cause discomfort and fatigue. These include sustained static neck flexion, shoulder flexion, forearm muscle exertion, extreme wrist postures, and prolonged standing (Lutz et al., 2001).

WWAL is associated with various well recognised health risks resulting from sustained exposure to the above and is a major contributing factor to WRMDs, such as carpal tunnel syndrome, tendonitis, thoracic outlet syndrome, and tension neck syndrome (Lutz et al., 2001). Each of these diagnostic terms is linked to certain types of occupational activity which affect various parts of the body resulting in these painful disorders.

A complete and useful understanding of the performance capabilities of workers on WWAL production lines requires knowledge of the mechanism of musculoskeletal dynamics. Thus, a brief explanation of this is presented in next section.

3.2 Investigation of musculoskeletal dynamics related walking and carrying

Motion such as walking and carrying is achieved by activation of the skeletal muscles (contracting and relaxing rhythmically), to produce the required kinetic energy. The activation of muscles causes bone loading and joint contact forces and consequently allows for moving the joints in a controlled fashion to accomplish the predetermined task requirements (Cappozzo, 1984).

Quite often, motions such as walking and carrying are influenced by a number of inter-individual factors, such as the weight and gender of worker (Brooks et al., 2005) as well as the effect of external forces such as the nature of the job requirements being undertaken (Cham & Redfern, 2004). In addition to these factors, the force-generation properties of the muscles, the anatomical features of the skeletal system (e.g. anthropometric properties, muscle paths) and the underlying neuronal control system, contribute substantially to generating the force to perform the tasks, such as supporting body weight, walking and carrying (LaFiandra et al., 2003).

4. Ergonomics measures in WWAL

In manual assembly systems, the focusing on only single aspects of ergonomics human performance measures may lead to conflicting conclusions in assessment of ergonomics stress level in work situations due to the following reasons (Al-Zuheri et al., 2010b):

- The possible interactions between more than one measure that may lead to conflicting conclusions about certain work hazards for the assemblers if these measures are considered separately,
- The large number of postures and the different exposures during manual assembly operations (as mentioned earlier) that should be considered in ergonomics evaluation, and
- The proposed ergonomically measures are sensitive to changes in the physical structure of workstations and workplaces in assembly systems.

Consequently, for obtaining accurate ergonomics understanding of work activities during manual assembly work, the evaluation process should examine by more than one measure to gain sufficiently precise data. The biomechanical and physiological measurements used have been instrumental in comparing different types of industrial jobs with respect to physical strain and fatigue (Garg et al., 1978; Bossi et al., 2004).

4.1 Ergonomics assessment of WWAL based on physiological and biomechanical models

4.1.1 Physiological model

Energy expenditure varies among assembly workers. The variations are caused by differing tasks involving work on components at various stages and walking from one place to another.

This is the most significant factor contributing to the variation of energy expenditure among assemblers (Honaker, 1996). Thus, average metabolic energy expenditure has been suggested for determining the amount of energy requirement needed to perform a given work without accumulating an excessive amount of physical fatigue (Garg et al., 1978).

Much research has been done estimating the energy expenditure of different assembly tasks (Holt et al., 1990; Chryssolouris et al., 2000; Ben-Gal & Bukchin, 2002; Longo et al., 2006). This research is used to ensure that the reasonable workload expectations are placed on the worker. This model can be used to estimate the energy expenditure of each task in WWAL; the parameters of the task being performed (e.g. object weight, speed, grade, and how a load is carried/moved in the hands/arms, height, etc.) as well as the individual factors such as gender, body weight and time taken to perform each task.

4.1.2 Biomechanical and dynamic motion models

Several biomechanical models have been developed to collect data on the nature of the strain placed on bodily structures and tissues by loads and forces during manual assembly processes (Kumar, 2006). The tools used to gather, and or analyse data in manual assembly works, included lifting limitations according to the National Institute for Occupational Safety and Health (NIOSH) guideline for biomechanical measure (Waters et al., 1994); workers posture during the task according to the Ovako Working-Posture Analysis System (OWAS) guidelines on risk or injury measure (Karhu & Kuorinka, 1977); cycle time from Methods Time Measurement (MTM) (Stevenson, 2002); and Rapid Upper Limb Assessment (RULA) (McAtamney & Corlett, 1993), is a measure for risk factors associated with upper limb disorders; Lifting Strength Rating (LSR) (Chaffin & Park, 1973); the university 3D Static Strength Prediction Program (3DSSPP) (Michigan, 2009); psychophysical approach (Snook & Hart, 1978); Lumbar Motion Monitor (LMM) and Ohio State University (OSU) Model (Davis & Waters, 1998).

Most of the mentioned biomechanical models are used to estimate the muscle forces in static postures. However, the effects of inertia are not accounted in these models; hence static models alone are not considered accurate enough to offer truly predictive data (Granata & Davis, 1999).

Much of the research undertaken on human dynamic motion, has been undertaken utilising the multi-segment models developed to assess moments of force or torque applied about the axes of the joints with the joint at various angles. Most of this research describes the biomechanical modelling of only one part of the body. A small proportion of that research has specifically addressed whole body models for activities involving both lower limbs and the upper body, such as whole body balance control (MacKinnon & Winter, 1993) and weight lifting (Kingma et al., 1996). However, none of this research has been focussed on biomechanical models that simulate dynamic walking and carrying conditions.

4.2 Suggested biomechanical model for the lower limbs of workers walking.

The worker in WWAL walks carrying work-pieces during movement from workstation to another sequentially (from workstation (1) to (2), to the point that the worker reaches the last one, workstation k), during the entire shift time. As stated earlier, this work is often associated with ergonomically poor conditions that result in WRMDs.

Therefore, there is a pressing need to propose a biomechanical model for effectively evaluating workers' capability to perform their required tasks without putting themselves at risk of developing a musculoskeletal injury.

In this research, a biomechanical model for the determination of net muscle moments and forces of lower limbs under dynamic motion conditions associated with performing assembly tasks of WWAL, in particular; walking and work-piece carrying. The resultant force and movements are calculated at the axis of hip, knee and ankle during level walking and carrying loads. In addition, the proposed model is used to investigate the possible effects of variables on the walking performance of workers during load-carriage tasks. These variables include walking speed and the weight of the work-piece carried.

Details of this model were fully described by Winter (Winter, 1980) and were validated by Flanagan and Salem (Flanagan & Salem, 2005) via comparing a top-down to a bottom-up study of squatting through measuring of net joint moments.

5. Materials and methods

5.1 The model: Net support moment approach

Biomechanical studies often total individual joint kinetics measurements (such as the net joint moment or net joint moment power) to obtain one biomechanical measurement of lower limb functions (Flanagan & Salem, 2005). However, it has been proposed that the net joint moments at the hip, knee, and ankle be collated into a single measure called "net support moment" (Winter, 1980).

The need for such a measure is justified (according to Winter), by the actual moments in the strength level required to walk, stand and recover from a slip etc. In addition, Winter found that some form of internal compensation was present. For example, when hip moment was high, knee moment or ankle moment was low, and vice versa. Consequently, interpreting the three moment curves in study as shown in figure 2, led him to suppose that the sum of all three moments (represent by support moment) plays a significant role in preventing a collapse of the knee.

Additionally to the above, Winter classified the joint moments to be positive when the pulling direction is counter clockwise and negative when clockwise, as shown in figure 3.

Equation 1 shows the net supporting moment calculated by summation of the three net joint moments (Winter, 1980):

$$M_s = M_k - M_a - M_h \tag{1}$$

Where M_s the net support moment, M_k, M_a and M_h are the moments at the knee, ankle and hip respectively. Assuming that the thigh and shank are equally long, the support moment was redefined by Hof (Hof, 2000). With same postulation of moment polarity, the new equation proposal by Hof is:

$$M_s = \tfrac{1}{2}M_a + M_k + \tfrac{1}{2}M_h \tag{2}$$

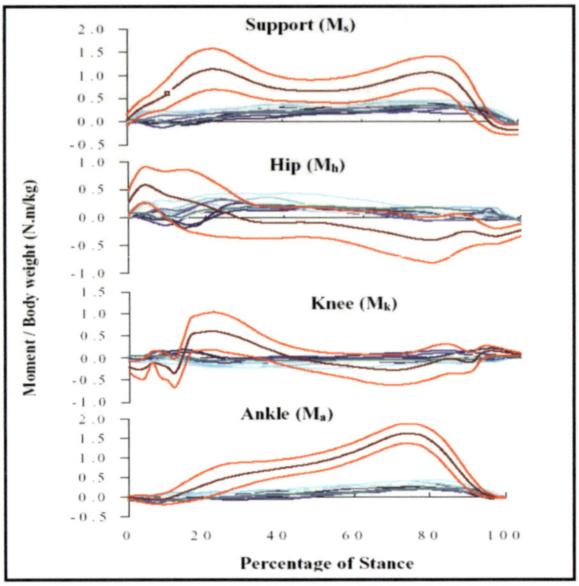

Fig. 2. Support and joint moments of force at hip, knee and ankle during walk (Winter, 2005).

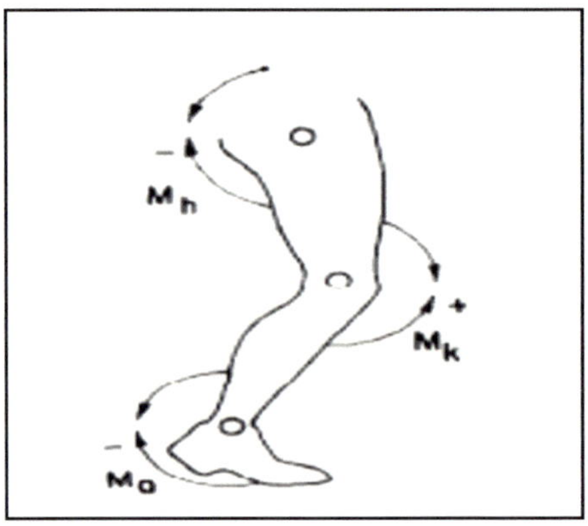

Fig. 3. Profile of joint moment of force at the ankle, knee and hip during walking (Winter, 2005).

This measure is commonly used for the assessment of mechanical output by lower limbs during walking (Winter, 1980; Hof, 2000), and in other activity such as sitting and standing (Yoshioka et al., 2009). While collating individual joint kinetic measures into a single measure (net support moment) has been used to characterize the mechanical demands of the lower limb across many activities, the validity of this single measure during dynamic occupational task like those in WWAL is still questionable.

Throughout this research, the goal of "net support moment" measure in an ergonomics context is to gain information about the overall mechanical demand placed on the muscles that cross each joint of lower limb. In other words, this measure is considered as the index for assessing the degree to which lower limb joints of the body are strained during manual tasks in WWAL.

5.1.1 Biomechanical modelling strategy

Biomechanical research uses laws of physics and engineering concepts to describe the motion undergone by the various body segments and forces during normal or abnormal activities. As a general approach, the human body is treated as a mechanical system, made up of rigid links (the bones) that are connected at joints. (Chaffin, 1969; Garg et al., 1982; Chaffin & Andersson 1990; Yanxin Zhang, 2005; Chaffin, 2007) have been presented a set of linked segment models of the human body that can be used to estimate forces and mechanical moments (torques) imposed on the system during work activities.

In these models, a part of the human body is modelled as a chain of rigid body segments, interconnected by joints. Intersegment reactive forces and moment loads at each joint of body member are calculated by applying Newton's second law and Euler's equations. Generally in Newton-Euler mechanics, the applied forces (i.e., body segment weights and hand loads) are multiplied by their perpendicular distance from joint centres (i.e. moment arms). Figure 4 illustrates many of the force and moment vectors at specific joints of the body including (hand, knee, elbow, ankle, shoulder, foot, and hip) can be calculated by the similar way (Michigan, 2009).

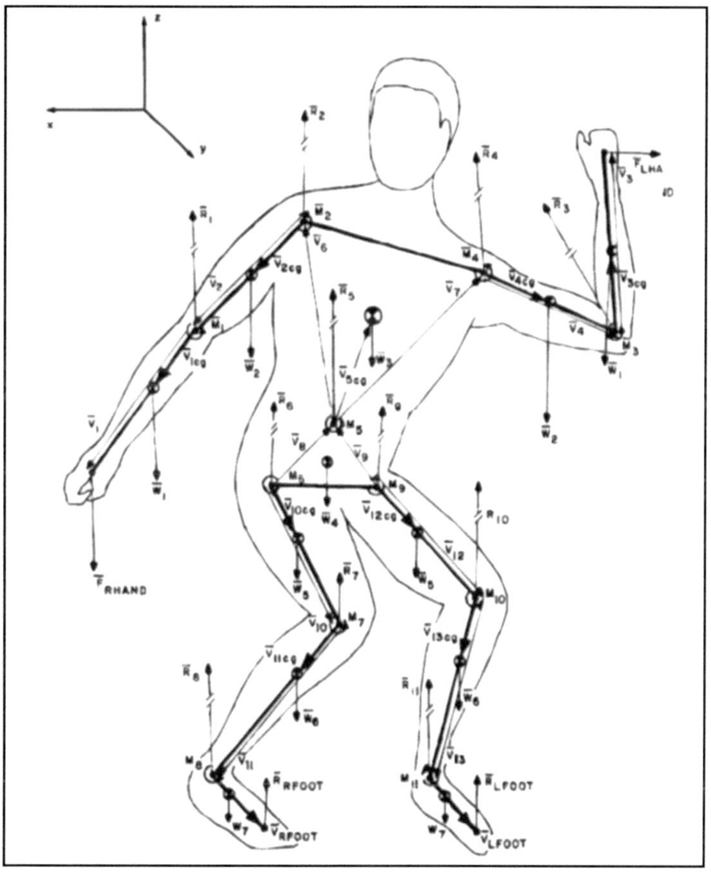

Fig. 4. Schematic representation of the strength model developed to calculate the muscle strength requirements needed to perform specified manual operations (Michigan, 2009).

Dynamic biomechanical analyses have been used in researches on walking and other activities such as lifting or carrying. In these analyses, inverse dynamics method is used to compute the joint moments of force in the lower limb (Redfern et al., 2001; Miller, 2002).

5.1.2 Newtonian model of the lower limb

The general logic that is used to predict forces and moments in lower limb joint during various jobs of WWAL is described in figure 5.

Accurate estimation of joint forces and moments of the lower limb during the occupational tasks of WWAL is mainly dependant on the accurate measurements for the static and inertial load during worker movement. The static load can be calculated by measure the following (Chaffin & Andersson, 1990; Wu & Ladin, 1996; Zijlstra & Bisseling, 2004):

1. Positions of the body segments, and
2. Foot-ground reaction force and moment.

While the calculation of inertial load due to requires kinematic description of the lower limb involves:

1. The position and orientation (joint angles of hip, knee and ankle), and
2. Walking speed and acceleration.

The above data describes the movement pattern (kinematic data) and the forces which cause that movement (kinetic data). Based on these data, an inverse dynamics method is applied in estimating the determinants of worker lower limb, such as the reaction forces in joints.

The method of inverse dynamics is used to derive the parameters of worker lower limb walking, starts from the foot segment toward the thigh segment with the motion data and the human body segmental characteristics that introduced in previous studies such as (Chaffin & Andersson, 1990; McLean et al., 2005).

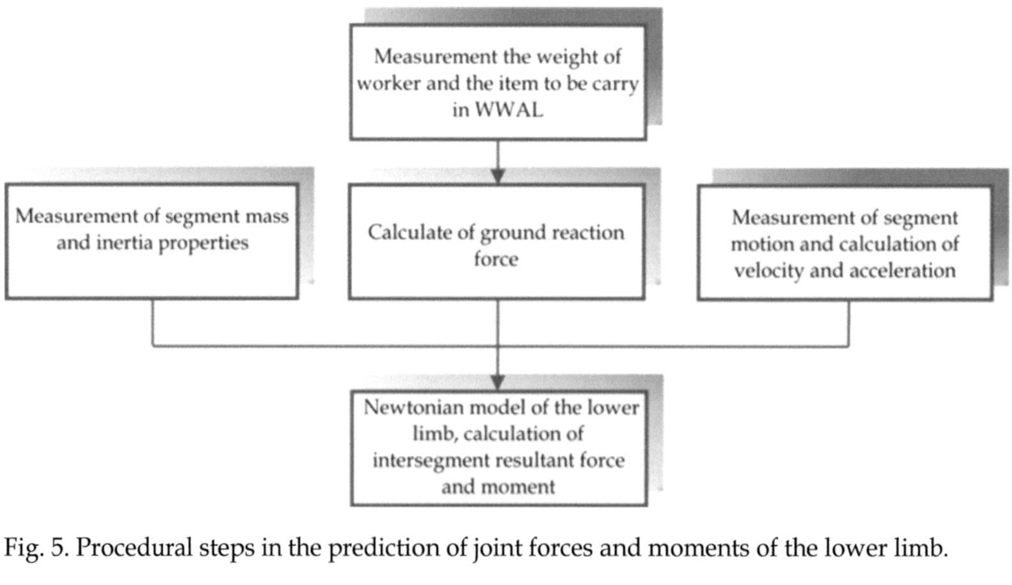

Fig. 5. Procedural steps in the prediction of joint forces and moments of the lower limb.

5.1.3 Assumptions

The model represents the movement of a lower limb of the human body. The three segment; foot, knee and thigh are treated as three rigid links as illustrated in figure 6. The joints included in the model are; ankle, knee and hip. Each leg has six Degrees of Freedom (DOF); three DOF at hip (flexion-extension, abduction-adduction and internal-external rotation), one DOF at knee (rotation about a fixed flexion-extension axis) and two DOF at ankle (rotation about talocrural and subtalar joint axes) (McLean et al., 2005). Given the weight of the work piece, inertial property of the segments, and length of the segments, the model is based on assumptions for appropriate approximations. These assumptions include:

1. The model for the sagittal plane, also can be applied in the frontal plane,
2. The model considers the two-handed asymmetric load -carrying during WWAL tasks,
3. The force of the load (the weight of work-piece to be assembled) passes through the central of mass of the hand,

The model is also based on several assumptions made regarding the muscle activity of the ankle, knee and hip, which follows in preceding studies of (Lin, 1995; Winter, 2005; Yanxin Zhang, 2005):

1. The centre of mass location of each segment remains fixed in the segment during the movement,
2. The worker body has not change in mass,
3. Throughout the movement, the length and cross-sectional area of each segment remains constant,
4. The joints are frictionless, and
5. Joints are considered to be hinge (2D motion) or ball and socket (3D motion).

The dimensions, mass, and internal properties of lower limb segments are assumed to conform to those proportions of anthropometric data provided in (Chaffin & Andersson, 1990; Winter, 2005).

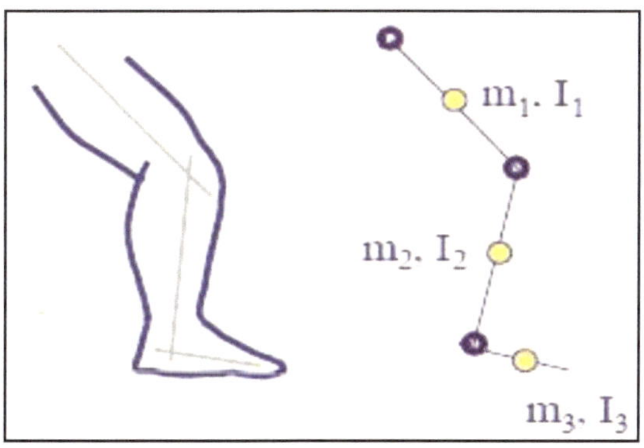

Fig. 6. A schematic representation of the lower limb segments where m_1, m_2 and m_3 and also I_1, I_2 and I_3 represent the mass and moment of inertia to the thigh, shank and foot respectively.

5.1.4 Model parameters

1. **Ground reaction force:** The forces which interact between the human foot and the ground in walking or running are referred to as Ground Reaction Force (GRF), as shown in figure 7. The GRF causes (Giddings et al. 2000):
   a. A forward acceleration on the body, and
   b. A moment about the vertical axis of body.

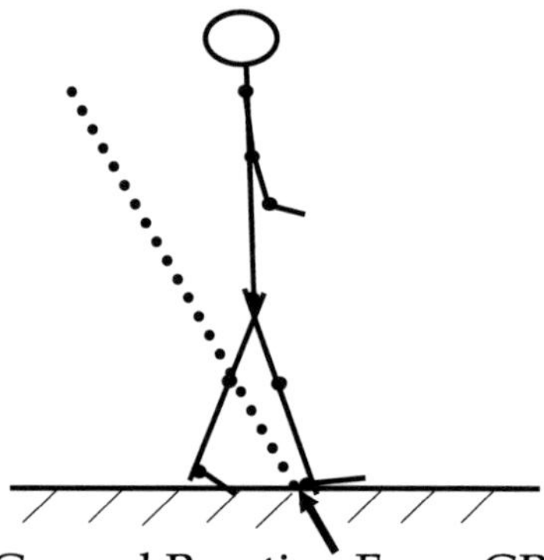

Fig. 7. Projection of GRF vector that is used to predicate the joint moments of force at the lower limb (reproduced from Winter, 2005).

The GRF can be calculated by using dynamic equations (Okada et al., 2006). The intersegment resultant forces and moments at the ankle, knee and hip are significantly dependent upon the magnitude of the GRF and its location relative to the joint centre for each. (Johnston et al., 1979). A number of researchers have examined GRF during walking (Redfern et al., 2001).

2. **Joints force and moment:** The control of walking is a result of the interaction of forces acting on human body. These forces can be internal or external. Internal forces refer to the inertial loads of the body segments which are related to the segmental acceleration. External forces on the body refer to gravitational and external loads (or static loads) due to the body contact with the environment (Wu & Ladin, 1996). In conclusion, the first one generates individual body segment movements, while the second affects whole body movements. The joint moments can be created by concentric and eccentric muscle contractions (Simonsen et al., 2007).
3. **Mass segments and inertia:** The inertia of the body segments is changed due to the non-uniform horizontal component of the propulsive force. Fluctuation in the amount of applied force will lead to change in the Centre of Gravity (COG) of the body segments. Variances in COG depend upon periods of speeding up and slowing down of the body segments (Cham & Redfern, 2004). Thus, the inertia of body segments changes also during the various activities of WWAL. The calculations of mass and inertial properties of each segment based on anthropometric measurements were made on the subjects as mentioned above.

The equations to calculate the joint forces and the moments and moment of inertia are described in next section.

5.1.5 Model formulation: Joint moments calculation

The drawing below (fig.8) is a Free-Body Diagram (FBD) representing the lower limb of a worker. The FBD demonstrates all the forces and moments that exist on the foot, shank and thigh. The equations derived to solve the resultant forces and moments are described below.

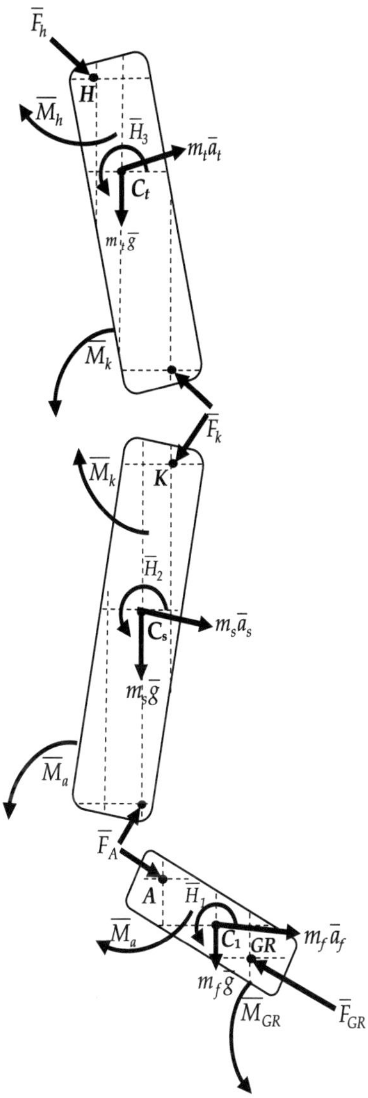

Fig. 8. The FBD of lower limb depicted with the intersegment resultant force and moment at hip, knee and ankle during walking.

Using inverse dynamics and the free body lower limb, much research concerned with the calculation of joint moments about the ankle, the knee and the hip joint (Hardt,

1978; Johnston et al., 1979; Wu & Ladin, 1996; Winter, 2005; Simonsen et al., 2007). This research uses slightly modified version of the formula presented by Johnston (Johnston et al., 1979). The resultants forces for the ankle, hip and knee are calculate as follows:

$$\overline{F}_a = m_1(\ddot{\overline{a}}_1 - \overline{g}) - \overline{F}_{GR} \tag{3}$$

$$\overline{F}_k = \sum_{i=1}^{2}\left[m_i(\ddot{\overline{a}}_i - \overline{g})\right] - \overline{F}_{GR} \tag{4}$$

$$\overline{F}_h = \sum_{i=1}^{3}\left[m_i(\ddot{\overline{a}}_i - \overline{g})\right] - \overline{F}_{GR} \tag{5}$$

Where:

$\overline{F}_{GR}$ = Ground reaction force acting on foot (kg),

$\overline{F}_j$ = Intersegment resultant force at the joint (j) (kg),

m_i =Mass of segment (i) (kg),

$\ddot{\overline{a}}_i$ =Acceleration vector of the centre of gravity of segment (i) (m/sec²), and

$\overline{g}$ = Acceleration vector due to gravity, 9.8 m/sec².

The sagittal plane joint moments that generated at the ankle, knee and hip can be computed using the following equations:

$$\overline{M}_a = \dot{\overline{H}}_1 - \left[\overline{r}_{A\_C_1} \times m_1(\ddot{\overline{a}}_1 - \overline{g})\right] - \overline{M}_{GR} - (\overline{r}_{GR\_A} \times \overline{F}_{GR}) \tag{6}$$

$$\overline{M}_k = \sum_{i=1}^{2}\left\{\dot{\overline{H}}_i - \left[\overline{r}_{K\_C_i} \times m_i(\ddot{\overline{a}}_i - \overline{g})\right]\right\} - \overline{M}_{GR} - (\overline{r}_{GR\_K} \times \overline{F}_{GR}) \tag{7}$$

$$\overline{M}_h = \sum_{i=1}^{3}\left\{\dot{\overline{H}}_i - \left[\overline{r}_{H\_C_i} \times m_i(\ddot{\overline{a}}_i - \overline{g})\right]\right\} - \overline{M}_{GR} - (\overline{r}_{GR\_H} \times \overline{F}_{GR}) \tag{8}$$

Where:

$\overline{M}_{GR}$ = The moment due to of the ground reaction force (kg),

$\overline{M}_i$ = Joint moment vector about joint (j),

C_i = Position of centre of mass of segment (i) (meter),

$\overline{r}_{j\_i}$ = Position vector from joint (j) to the centre of gravity of segment (i) (meter), and

$\dot{\overline{H}}_i$ = Inertial component vector of the joint moment about joint (j) (kg.m²).

The moment of inertia about the pivot point of joint (j) can be calculated by using the following equation derived from the basic mechanics:

$$\dot{\bar{H}}_i = I_i \times \bar{\alpha}_i \tag{9}$$

Where:

I_i = Moment of inertia of segment (i) about the centre of mass (kg.m²), and

$\bar{\alpha}_i$ =Angular acceleration vector of segment (i) about the centre of mass (rad. / s²).

The following points were taken as positions for the lower limb joints centres:

A = Position of ankle joint centre (meter),
K = Position of hip knee centre (meter),
H = Position of hip joint centre (meter), and
GR = Position of ground reaction force effect (meter).

5.2 The hypothetical assembly line

This research considers a U-shaped manual assembly line using walking worker approach and multifunctional workers to assemble a single model product (hydraulic valve actuator). The line is depicted in figure 9. The weight of hydraulic valve actuators that is assembled and handled manually at first workstation is 4.96 kg. Table 1 summarizes the example at each workstation in terms of the weight of the valve actuator after assembly process at each workstation on the line.

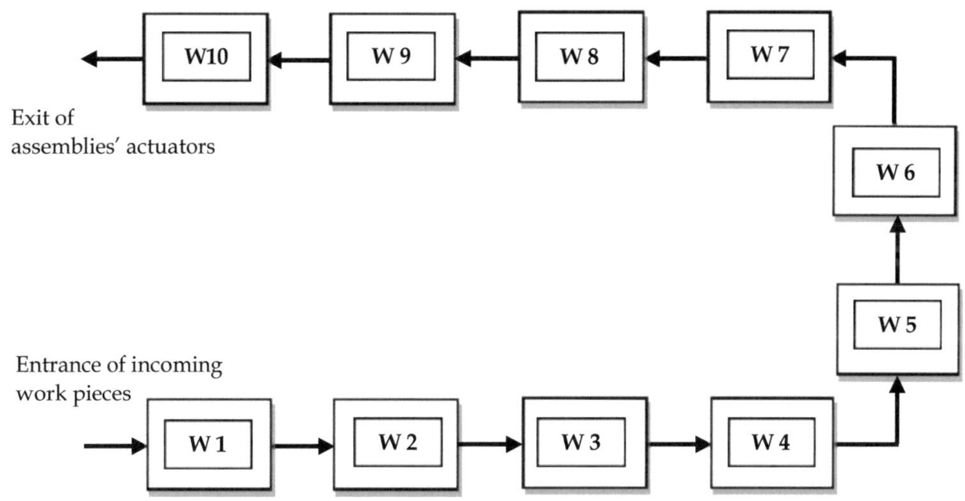

Fig. 9. U-shaped assembly line of hydraulic valve actuators for *(m)* workers and *(k)* workstations.

The work-piece is processed at workstations w1, w2,. . . , w10, sequentially, and departs from the line as a finished product. The worker arrives at workstation w (any workstation on the line) to perform the specified processing of the part assemblies at that workstation. When the operation at workstation w is accomplished, then he walks toward next workstation and carries out the essential assembly work as scheduled and continues until the product is built completely at workstation w10. The worker then moves back to the first workstation w1 to begin the assembly of a new product with the same previously described procedure. The products being assembled are transported manually by workers between workstations of assembly line.

Workstations	**Weight of actuators after process (Kg)**
k 1	4.96
k 2	5.70
k 3	6.27
k 4	7.22
k 5	7.58
k 6	7.96
k 7	8.750
k 8	9.63
k 9	9.63
k 10	9.63
	Final weight of actuator = 9.63Kg

Table 1. The weight of the actuator (the work-piece) at each workstation on the line

5.3 Experimental data and procedure

Part of this research was performed based on previously published data on the calculation of net summation of the moments at three joints (hip, knee and ankle), support moment (Winter, 1980; Hof, 2000). That data includes; (1) segmental relative weight, centre of gravity and moment of inertia data for the (hypothetical) workers as shown in table 2; (2) the resolution of the position of the body from the angles at each articulation; (3) the determination of the angular velocities and angular accelerations at each articulation, which in turn, gives the linear acceleration of the body links.

It is based upon the assumption that the average weight of workers is 82 kg. The procedure of modelling includes two stages. Firstly, the calculation stage; this consists of several steps; (1) the calculation of inertial forces and inertial resistance moments due to acceleration: (2) calculation of moments and forces on the body from the motion input data (i.e. the x-y joint position data over time for the ankle, knee and hip); (3) the calculation of reactive moments and forces at each articulation exerted by the muscles to overcome the resultant forces due to external loads and body weight; and (4) the joint moments of all lower limb joint moments (hip, knee, and ankle) and also support moment were calculated.

In the second stage, on the basis of inputting walking speed and the weight of the work piece, the effect of these variables at lower limbs joints is estimated. The model application consisted of using this data with two walking speed with carrying work-piece to be assembled; (1) slow walking (0.7 meter/sec.) and (2) fast walking (1.4 meter/sec.). The initial weight of work-piece is 4.96 kg, increasing gradually with assembling process to reaching a final weight of 9.63 kg at workstations 8, 9 and 10. The carrying technique is front with two-hand.

Segment	Relative Weight	Centre of Gravity	Moment of Inertia about CG (kg.m²)
Head	0.073	46.4% vertex to chin	0.0248
Trunk	0.507	38.0% shoulder to hip	1.2606
Upper Arm	0.026	51.3% shoulder to elbow	0.0213
Forearm	0.016	39.0% elbow to wrist	0.0076
Hand	0.007	82.0% wrist to knuckle	0.0005
Thigh	0.103	37.2% hip to knee	0.1052
Calf	0.043	37.1% knee to ankle	0.0504
Foot	0.015	44.9% heel to toe	0.0038

Table 2. Segmental relative weight, centre of gravity and moment of inertia data

6. Results and discussion

6.1 Joint moments

For the stance phase normalized to 100%, figures (10-b, 10-c and 10-d) represent the lower limb joint moments (hip, knee and ankle joints) on the sagittal plane for a single worker in WWAL during complete posture cycle under both normal walking and different work-piece carrying conditions. As illustrated in that figure, changes in the relative shape and magnitude were found in moment of hip, knee and ankle joints among different weights carrying (4.96, and 9.63 kg) and also basically when workers walking without carrying any work-piece.

Among the three lower limb joints, the hip moment, which was consistently and significantly more biased with increasing work-piece weight during the 0-10% of the gait cycle. This can be explained by the fact that the positive hip joint angular impulse for the contralateral side tended to increase with the increase of work-piece weight.

From the results of this research, it was found that when workers walked at different speeds carrying work- pieces, the moment of the lower limb joint would increase with the walking speed (figure 11). This is because the percentage of the stance phase decreased as the walking speed increased and the swing phase increased as the walking speed increased. Figure 10 presents the M_S and the contributions to the M_S of each joint for the stance phase normalized to 100%.

6.2 Net support moment

Figure 10 presents the contributions of each lower limb joint for the net support moment in carrying different work-piece weights as well as in normal walking. At the initial stage of the gait cycle (0-10%), the hip and knee joint moments were large. On the other hand, during that stage the ankle joint moment was nearly zero. Therefore, the hip and ankle contributed to the most part of the support moment throughout the stance phase.

Net support moment was considerably reduced and even in negative values throughout the swing phase (50–100% of the gait cycle). This is because of the negative values of the hip moment in end of stance phase. From approximately 40% to 100% of the gait cycle, the knee and the ankle moment contributed positively to support the body-mass of worker during the job.

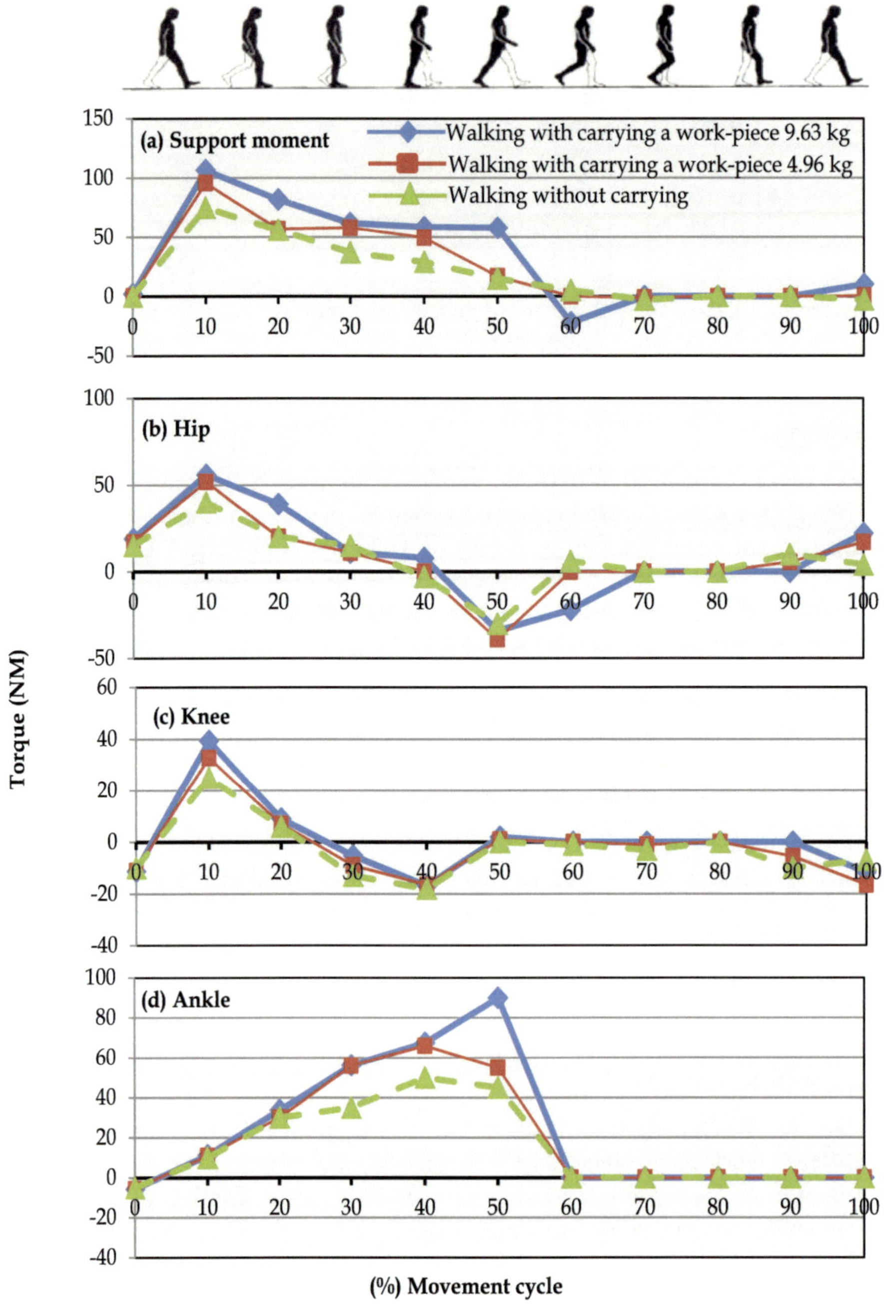

Fig. 10. The calculated net support moment M_S according to the original definition of Winter and the resulting moment about the (b) hip moment; (c) knee moment and (d) ankle moment. The calculations were performed for different loading condition (as shown in legend). Workers data: male, 34 years, weight 82 kg and walking speed 0.7 m/sec.

From figure 11, it was evident, that support moment varies quite considerably when walking speed was increased from 0.7 to 1.4 meter/sec. This can be explained by the fact that both the GRF and the knee angle in stance are strongly dependent on walking speed. As expected, there was a significant increase in net support moment throughout the walking from workstation to another (figure 11). However, this was related significantly to increasing weight of work-pieces during assembly due to the addition of new components at to the work-piece at each workstation.

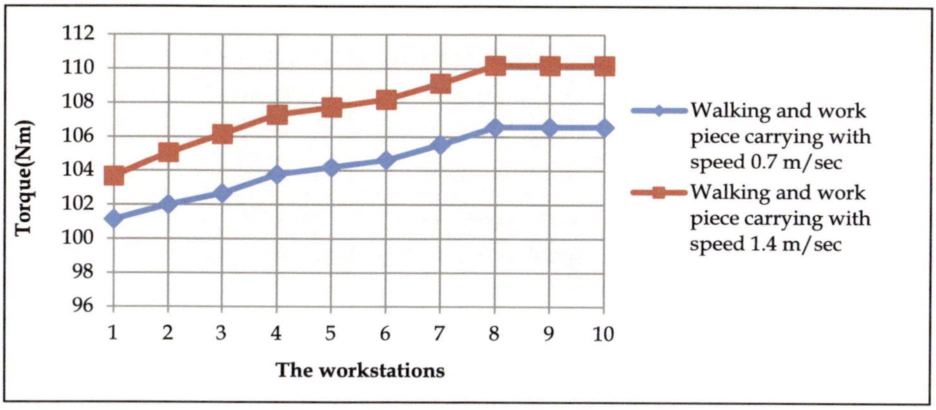

Fig. 11. Net support moment M_S at WWAL workstations, for two different carrying condition during the 10% stance period of movement cycle where the M_S at that time reaches to the maximum value as the results indicated in figure 10.

6.3 Ground reaction forces

Ground reaction forces in walking increased significantly with work-piece carrying (figure 12). More specifically, carrying a 4.96 kg weight of work-pieces at first workstation and then increased to 9.63 kg load led to increases in the peak normal ground reaction force ranging from -75 to -50 N and to -50 from the normal walking, respectively.

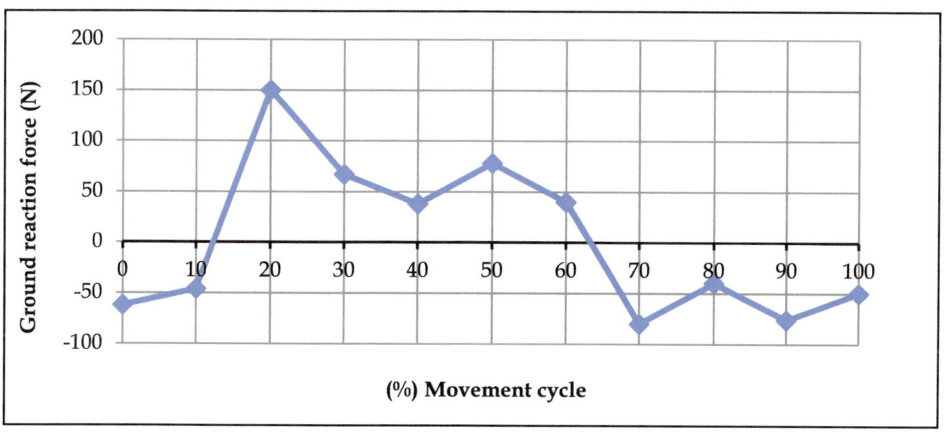

Fig. 12. Calculated ground reaction force during complete movement cycle (0-100%) while walking with carrying 4.96 weight for work-piece.

Normally, ground reaction forces depend on body mechanics, mass, and acceleration at the time when the individual touches the ground. Consequently, as mass (weight of work-piece) increases, ground reaction forces generally increases. Also, a worker's gait pattern affects ground reaction forces. As a result, intensity, mass, speed and type of activity were expected to be significant fixed effects.

7. Conclusion and further work

The net support moment model, described by Winter in 1980, provides a useful framework to study the strategy used to support body weight during walking while performing a job in dynamic production systems like WWAL. In this research, the model was used to predict the moments for the hip, knee and ankle during walking and carrying different work-piece weights, as well as normal walking. In addition to work-piece weight, the effect of walking speed of walker on support moment was also investigated. In conclusion, the results of this research indicated that, the net support moment and the contributions the hip, knee and ankle moment respectively, is an interesting method to assess the weight bearing and walking speed strategies for walking workers.

The net support moment is calculated by summation of the moments at hip, knee and ankle during walking. This enables a designer to construct a layout of WWAL in such a way as to obtain optimal movement, i.e. the movement in which the net support moment of all three joints is minimized.

The reliability of the presented predictive model as a tool to investigate the mechanical demands of the lower limbs during dynamic occupational task like those in WWAL is still questionable. To enable the use of this model, it needs further work in two areas; (1) model validation by comparing the predicted results with the actual measurement data for dynamic walking whilst carrying a work- piece, (2) further investigation of the relationships of other variables during walking of worker with work-piece carriage, such as gender and the physical design of workstations.

8. Acknowledgment

The authors would like to express their appreciation to anonymous referees for their helpful comments.

9. References

Al-Zuheri, A.; Luong, L. & Xing, K. (2010a). An Integrated Design Support Methodology for Walking Worker Assembly Lines, *Proceedings of IMECS 2010 International Multi-Conference of Engineers and Computer Scientists*, pp. 1612-1617, ISBN 978-988-17012-8-2, Hong Kong, March 17-19, 2010

Al-Zuheri, A.; Luong, L. & Xing, K. (2010b). Ergonomics Design Measures in Manual Assembly Work, *Proceedings of 2nd ICESMA 2010 International Conference on Engineering Systems Management and Applications*, pp. 53-58, ISBN 978-1-4244-6520-0, Sharjah, UAE, March 30 –April 1, 2010

Ben-Gal, I. & Bukchin, J. (2002). The Ergonomic Design of Workstations Using Virtual Manufacturing and Response Surface Methodology, *IE Transactions,* Vol. 34, No. 4, (April 2002), pp. 375-391, ISSN 0740-817X

Bischak, DP. (1996). Performance of a Manufacturing Module with Moving Workers, *IIE Transactions,* Vol. 28, No. 9, (September 1996), pp. 723-733, ISSN 0740-817X

Bley, H.; Zenner, C. & Bossmann, M. (2007), Walking Worker Assembly Lines - A Contribution to Lean Production. *Proceedings COMA 2007 International Conference Competitive Manufacturing - The Challenge of Digital Manufacturing,* pp. 297-302, Stellenbosch, South Africa, 31 January - 2 February, 2007

Bossi, LL.; Stevenson, JM.; Bryant, JT.; Pelot, RP.; Reid, SA. & Morin, EL. (2004). Development of a Suite of Objective Biomechanical Measurement Tools for Personal Load Carriage System Assessment, *Ergonomics,* Vol. 47, No. 12, (October 2004), pp. 1255 - 1271, ISSN 0014-0139

Brooks, AG.; Gunn, SM.; Withers, RT.; Gore, CJ. & Plummer, JL. (2005). Predicting Walking Mets and Energy Expenditure from Speed or Accelerometry, *Medicine and Science in Sports and Exercise,* Vol. 37, No. 7, (July 2004), pp. 1216-1223

Cappozzo, A. (1984). Gait Analysis Methodology, *Human Movement Science,* Vol. 3, No. 1-2, (March 1984), pp. 27-50

Chaffin, D. & Park, K. (1973). A Longitudinal Study of Low-Back Pain as Associated with Occupational Weight Lifting Factors, *American Industrial Hygiene Association Journal,* Vol. 34, No. 12, (December 1973), pp. 513–25, ISSN 0002-8894

Chaffin, DB. (1969). A Computerized Biomechanical Model - Development of and Use in Studying Gross Body Actions, *Journal of Biomechanics,* Vol. 2, No. 4, (October 1969), pp. 429-441

Chaffin, DB. (2007). Human Motion Simulation for Vehicle and Workplace Design. *Human Factors and Ergonomics in Manufacturing,* Vol. 17, No. 5, (September 2007), pp. 475-484, ISSN 1090-8471

Chaffin, DB. & Andersson, GBJ. (1990). *Occupational Biomechanics* (2nd Edition), John Wiley & Sons, ISBN 978-0-471-72343-1, New York, USA

Cham, Re & Redfern, MS. (2004). Gait Adaptations During Load Carrying on Level and Inclined Surfaces, *Occupational Ergonomics,* Vol. 4, No. 1, (May 2004), pp. 11–26, ISSN 1359-9364

Chryssolouris, G.; Mavrikios, D.; Fragos, D. & Karabatsou, V. (2000). A Virtual Reality-Based Experimentation Environment for The Verification of Human-Related Factors in Assembly Processes, *Robotics and Computer-Integrated Manufacturing,* Vol. 16, No. 4, (August 2000), pp. 267-276, ISSN 0736-5845

Davis, KG.; Marras, WS. & Waters, TR. (1998). Evaluation of Spinal Loading During Lowering and Lifting, *Clinical Biomechanics,* Vol. 13, No. 3, (April 1998), pp. 141-152, ISSN 0268-0033

Flanagan, SP. & Salem, GJ. (2005). The Validity of Summing Lower Extremity Individual Joint Kinetic Measures, *Journal of Applied Biomechanics,* Vol. 21, (May 2005), pp. 181-188, ISSN 1065-8483

Garg, A.; Chaffin, DB. & Herrin, G. (1978). Prediction of Metabolic Rates for Manual Materials Handling Jobs, *American Industrial Hygiene Association Journal,* Vol. 39, No. 8, (August 1978), pp. 666-764, ISSN 0002-8894

Garg, A.; Chaffin, DB. & Freivalds, A. (1982). Biomechanical Stresses From Manual Load Lifting - A Static vs Dynamic Evaluation, *IIE Transactions,* Vol. 14, No. 4, pp. 272-281, ISSN 0569-5554

Giddings, VL.; E, GSB.; Whalen, RT. & Carter, DR. (2000). Calcaneal Loading During Walking and Running, *Medicine and Science in Sports and Exercise,* Vol. 32, No. 3, (March 2000), pp. 627-634, ISSN 0195-9131

Granata, K.; Marras, W. & Davis, K. (1999). Variation in Spinal Load and Trunk Dynamics During Repeated Lifting Exertions, *Clinical Biomechanics,* Vol. 14, No. 6, (July 1999), pp. 367-375, ISSN 0268-0033

Grassi, A.; Gamberi, M.; Manzini, R. & Regattieti, A. (2004). U -Shaped Assembly Lines with Stochastic Tasks Execution Times: Heuristic Procedures for Balancing and Re-Balancing Problems, *Proceedings of SCS 2004 Society for Modeling and Simulation International,* pp. 137-143, ISSN 0735-9276, Arlington, Virginia, USA, April 18-22, 2004

Hardt, DE. (1978). Determining Muscle Forces in The Leg During Normal Human Walking-an Application and Evaluation of Optimization Methods, *Journal of Biomechanical Engineering,* Vol. 100, (May 1978), pp. 72-78

Hof, AL. (2000). On The Interpretation of The Support Moment, *Gait and Posture,* Vol. 12, No. 3, (December 2000), pp. 196–199, ISSN 0966-6362

Holt, KG.; Hamill, J. & Andres, RO. (1991). Predicating The Minimal Energy Costs During Human Walking, *Medicine and Science in Sport and Exercise,* Vol. 23, No. 4, (April 1991), pp. 491-498

Honaker, RE. (1996). Assessing Trailer Material Handling Tasks: Biomechanical Modelling, Posture Categorization, Physiological Measure, and Subjective Rating, Industrial and Systems Engineering, Virginia Polytechnic Institute and State University

Hunter, SL. (2002). Ergonomic Evaluation of Manufacturing System Designs, *Journal of Manufacturing Systems,* Vol. 20, No. 6, (December 2002), pp. 429-444, ISSN 0278-6125

Johnston, RC.; Brand, RA. & Crowninshield, RD. (1979). Reconstruction of The Hip. A Mathematical Approach to Determine Optimum Geometric Relationships, *The Journal of Bone and Joint Surgery,* Vol. 61, No. 5, (July 1979), pp. 639-652, ISSN 0021-9355

Karhu, O.; P, K. & Kuorinka, I. (1977). Correcting Postures in Industry: a Practical Method for Analysis, *Applied Ergonomics,* Vol. 8, No. 4, (December 1977), pp. 199-201

Kingma, I.; Looze, MPD., Toussaint, HM., Klijnsma, HG. & Bruijnen, TBM. (1996). Validation of a Full Body 3-D Dynamic Linked Segment Model, *Human Movement Science,* Vol. 15, No. 6, (December 1996), pp. 833-860, ISSN 0167-9457

Kumar, R. (2006). Ergonomic Evaluation and Design of Tools in Cleaning Occupation, Department of Human Work Sciences, Luleå University of Technology

LaFiandra, M.; Wagenaar, RC.; Holt, KG. & Obusek, JP. (2003). How Do Load Carriage and Walking Speed Influence Trunk Coordination and Stride Parameters?, *Journal of Biomechanics,* Vol. 36, No. 1, (January 2003), pp. 87–95, ISSN 0021-9290

Lin, CJ. (1995). A Computerized Dynamic Biomechanical Simulation Model for Sagittal Plane Lifting Activities, Industrial Engineering, Texas Tech University

Longo, F.; Mirabelli, G. & Papoff, E. (2006). Effective Design of an Assembly Line Using Modeling & Simulation, *Proceedings of WSC 2006 38th Winter Simulation*, pp. 1893-1898, ISBN 1-4244-0501-7, Monterey, CA, USA , December 03 - 06, 2006

Lutz, TJ.; Starr, H.; Smith, CA.; Aaron M. Stewart; Monroe, MJ.; Joines, SMB. & Mirka, GA. (2001). The Use of Mirrors During an Assembly Task: A Study of Ergonomics and Productivity, *Ergonomics*, Vol. 44, No. 2, (January 2001), pp. 215 - 228, ISSN 1071-1813

MacKinnon, CD. & Winter, DA. (1993). Control of Whole Body Balance in The Frontal Plane During Human Walking, *Journal of Biomechanics*, Vol. 26, No. 6, (June 1993), pp. 633-644

McAtamney, L. & Corlett, EN. (1993). RULA: A Survey Method for The Investigation of World-Related Upper Limb Disorders, *Applied Ergonomics*, Vol. 24, No. 2, (April 1993), pp. 91-99

McLean, SG.; Su, A. & Bogert, Ajvd. (2005). Development and Validation of A 3-D Model to Predict Knee Joint Loading During Dynamic Movement, *Journal of Biomechanical Engineering*, Vol. 125, No. 6, (December 2003), pp. 864-873

Melin, B.; Lundberg, U.; Derlund, JS. & Granqvist, M. (1999). Psychological and Physiological Stress Reactions of Male and Female Assembly Workers: A Comparison Between Two Different Forms of Work Organization, *Journal of Organizational Behavior*, Vol. 20, No. 1, pp. 47-61, ISSN 0894-3796

Michigan, Uo. (2009). *3D Static Strength Prediction Program Version 6.0.2 User's Manual*, Center for Ergonomics, Ann Arbor

Mileham, A.; Jeffries, A.; Owen, G. & Pellegrin, F. (2000). The Design and Performance of Walking Fitter Assembly Lines, *Proceedings of CARS&F 2000 16th International Conference on CADCAM Robotics and Factories of the Future*, pp. 325-332, University of the West Indies, Trinidad

Miller, D. (July 2002). A Method to Determine The Forces on and The Movement of Joints and Segments of The Human Body, In University of Waterloo, 20.07.2010, Available from <http://stargate.uwaterloo.ca/~jafoster/SOTAReport>

Miller, J. (December 2007). Toyota's Suggestion System: 56 Years and Still Going Strong, 20.01.2011, Available from <http://www.gemba.com/consulting.cfm>

Miltenburg, J. (2001). U-shaped Production Lines: A Review of Theory and Practice, *International Journal of Production Economics*, Vol. 70, No. 3, (April 2001), pp. 201-214, ISSN 0925-5273

Miyake, DI. (2006). The Shift From Belt Conveyor Line to Work-Cell Based Assembly Systems to Cope With Increasing Demand Variation and Fluctuation in The Japanese Electronics Industries, *International Journal of Automotive Technology and Management*, Vol. 6, No. 4, (December 2006), pp. 419-439

Moller, T.; Mathiassen, SE.; Franzon, H. & Kihlberg, S. (2004). Job Enlargement and Mechanical Exposure Variability in Cyclic Assembly Work, *Ergonomics*, Vol. 47, No. 1, (January 2004), pp. 19-40, ISSN 0014-0139

Nakade, K. & Nishiwaki, R. (2008). Optimal Allocation of Heterogeneous Workers in A U-Shaped Production Line, *Computers & Industrial Engineering*, Vol. 54, No. 3, (April 2008), pp. 432-440, ISSN 0360-8352

Okada, H.; Ae, M. & Robertson, DGE. (2006). *The Effect of Ground Reaction Force Components on Ankle Joint Torque During Walking*, University of Ottawa, Canada, Ottawa

Redfern, MS.; Cham, R.; Gielo-Perczak, K.; Nqvist, RG.; Hirvonen, M.; Lanshammar, HK.; Marpet, M. & Pai, CY-C (2001). Biomechanics of Slips, *Ergonomics*, Vol. 44, No. 13, (November 2010), pp. 1138 - 1166, ISSN 0014-0139

Simonsen, EB.; Dyhre-Poulsen, P.; Voigt, M.; Aagaard, P. & Fallentins, N. (2007). Mechanisms Contributing to Different Joint Moments Observed During Human Walking, *Scandinavian Journal of Medicine & Science in Sports*, Vol. 7, No. 1, (January 2007), pp. 1-13, ISSN 0905-7188

Snook, SH.; Campanelli, RA. & Hart, JW. (1978). Study of Three Preventive Approaches to Low Back Pain, *Journal of Occupational Medicine*, Vol. 20 No. 7, (July 1978), pp. 478-481

Stevenson, WJ. (2002). *Operations Management*, (7th Edition), McGraw-Hill Irwin, ISBN 0073377848, New York, USA

Stockton, DJ.; Ardon-Finch, J. & Khalil, R. (2005). Walk Cycle Design for Flexible Manpower Lines Using Genetic Algorithms, *International Journal of Computer Integrated Manufacturing*, Vol. 18, No. 1, (January-February 2005), pp. 15 - 26, ISSN 0951-192X

Wang, Q.; Lassalle, S.; Mileham, AR. & Owen, GW. (2009). *Analysis of a Linear Walking Worker Line Using a Combination of Computer Simulation and Mathematical Modeling Approaches*, University of Bath, Bath, UK

Wang, Q.; Owen, G. & Mileham, A. (2005). Comparison Between Fixed- and Walking-Worker Assembly Lines, *Proceedings of the Institution of Mechanical Engineers, Part B: Journal of Engineering Manufacture*, Vol. 219, No. 11, (November 2005), pp.845-848, ISSN 0954-4054

Waters, T.; Putz-Anderson, V. & Garg, A. (1994). *Applications Manual for the Revised NIOSH Lifting Equation*, National Institute for Occupational Safety and Health, Cincinnati, Ohio

Winter, DA. (2005). *Biomechanics and Motor Control of Human Movement*, (3rd edition), ISBN 978-0-470-39818-0, Wiley-Interscience, Hoboken, New Jersey

Winter, DA. (1980). Overall Principle of Lower Limb Support During Stance Phase of Gait, *Journal of Biomechanics*, Vol. 13, No. 11, (November 1980), pp. 923-927

Wu, G. & Ladin, Z. (1996). Limitations of Quasi-Static Estimation of Human Joint Loading During Locomotion, *Medical and Biological Engineering and Computing*, Vol. 34, No. 6, (November 1996), pp. 472-476

Yanxin Zhang, BS. (2005). *3D Simulation of Manual Material Handling Tasks Based on Nonlinear Optimization Method*, Industrial Engineering, Texas Tech University, Texas, USA

Yoshioka, S.; Nagano, A.; Hay, DC. & Fukashiro, S. (2009). Computation of The Kinematics And The Minimum Peak Joint Moments of Sit-to-Stand Movements, *BioMedical Engineering OnLine*, Vol. 6, No. 1, (July 2007), pp. 1-14, ISSN 1475925X

Zavadlav, E.; McClain, JO. & Thomas, LJ. (1996). Self-Buffering, Self-Balancing, Self-Flushing Production Lines, *Management Science*, Vol. 42, No. 8, (August 1996), pp. 1151-1164

Zijlstra, W. & Bisseling, R. (2004). Estimation of Hip Abduction Moment Based on Body Fixed Sensors, *Clinical biomechanics (Bristol)*, Vol. 19, No. 8, (October 2004), pp. 819–827, ISSN 0268-0033

7

Usability of Interfaces

Mário Simões-Marques[1] and Isabel L. Nunes[2]
[1]Portuguese Navy,
[2]Centre of Technologies and Systems,
Faculdade de Ciências e Tecnologia, Universidade Nova de Lisboa,
Portugal

1. Introduction

In recent years human society evolved from the "industrial society age" and transitioned into the "knowledge society age". This means that knowledge media support migrated from "pen and paper" to computer-based Information Systems.

This evolution introduced some technological, organizational, and methodological changes affecting the demand, workload and stress over the workers, many times in a negative way. Due to this fact Ergonomics assumed an increasing importance, as a science/technology that deals with the problem of adapting the work to the man, namely in terms of usability.

Usability is a quality or characteristic of a product that denotes how easy this product is to learn and to use (Dillon, 2001); but it is also an ergonomic approach, and a group of principles and techniques aimed at designing usable and accessible products, based on user-centred design.

User-centred design is a structured development methodology that focuses on the needs and characteristics of users, and should be applied from the beginning of the development process in order to make software applications more useful and easy to use (Averboukh, 2001; Nunes, 2006).

Formally, usability is defined as "the extent to which a product can be used by specified users to achieve specified goals with effectiveness, efficiency and satisfaction in a specified context of use" (ISO 9241 - Part 11) (ISO 9241, 1998). Therefore usability is a relative concept, which is dependent on several factors. It applies equally both to hardware and software design.

Adequate usability is important because it is a characteristic of product quality that leads to improving product acceptability, increasing user satisfaction, improving product reliability and it is also financially beneficial to companies. Such benefit can be seen from two points of view, one related with workers' productivity (less training time and faster task completion), and the other with product sells (products are easier to sell and market themselves, when users had positive experiences) (Nunes, 2006).

A product designed with the user psychological and physiological characteristics in mind, is more efficient to use (less time to accomplish a particular task), easier to learn (operations can be learned by observing the object), and more satisfying to use (Nielsen, 1993).

Usability covers a broad spectrum of aspects regarding a product. Goud (Gould, 1995) argues that, despite several of these aspects are least discussed in literature, usability of components include components as System performance, System functions, User interface, Reading materials, Language translation, Outreach program, Ability for costumers to modify and extend, Installation, Field maintenance and serviceability, Advertising or Support-group users. However, some authors see this broad spectrum of issues as a step beyond usability, which is designated as User Experience Design. This theme is addressed in another chapter of the present book. The present chapter addresses usability in a traditional way, the one that relates mainly with the usability of interfaces, including aspects of system performance and system functions.

Literature describes a number of methodologies and tools that contribute to ensure the usability of a product considering, namely, the phase of development in which they are applied, the amount of resources they require and how they are applied (e.g., synthesized in (Usability Partners, 2011)). Most of these tools or methods are dedicated to a specific phase of project development (design phase, preliminary design and prototyping phase, and test and evaluation phase), some are applied in two different phases, and very few are appropriate to be applied in the three phases. In this chapter we will discuss with some detail the test and evaluation phase considering different methods, such as, analytic and heuristic evaluations, and SUMI.

A quite new challenge in terms of usability is the context of use of applications that exploit new forms of interfacing with the product, such as the use of touch and multitouch interfaces. The body of knowledge available is still limited, nevertheless, there is a vast literature on guidelines and good practices for generic usability, which can also be adapted to the context of touch and multitouch interfaces (e.g., (Microsoft, 2009), (MSDN, 2011), (HHS, 2006), (Largillier et al., 2010), (Meador et al., 2010), (Kreitzberg & Little, 2009), (Capra, 2006)).

The present chapter presents an overview of the general principles to observe when a user-centred design is adopted, provides a summary of methods and tools that are available to support the design and evaluation of products with good usability, offers examples of guidelines and good practices that can be adopted.

2. Usability and interfaces – Basic principles and heuristics

In some countries usability is a legal obligation. For instance, in European Union according to the Council Directive, 90/270/EEC, of 29 May, on the minimum safety and health requirements for work with display screen equipment, when designing, selecting, commissioning and modifying software the employer shall take into account the following principles:

- The software must be suitable for the task;
- The software must be easy to use and adaptable to the operator's level of knowledge or experience;
- Systems should provide users with information on its operation;
- Systems must display information in a format and at a pace adapted to users;

- The principles of software ergonomics must be applied, in particular to human data processing.

Therefore to meet these requirements the software development should be accompanied by an evaluation of its usability.

In simple terms, the usability of a system can be seen as the ease with which the system is used by its users, i.e., with the characteristic of being easy to use, or as is often said, to be "user friendly".

Therefore, usability is a feature of interaction between the user and the system. Usability evaluation can be based on a set of attributes, such as, operator performance (completing a task with reduced turnaround times and low error rates), satisfaction or ease of learning.

Usability can also be seen as synonymous of product quality, namely of software quality.

Usability is a critical aspect to consider in the development cycle of applications which requires a user-centred design and carrying out usability testing. Such tests cannot ignore the context of use of the software, which is essential to conduct usability studies.

When human-machine interfaces are built taking into account usability criteria, interfaces are capable of allowing an intuitive, efficient, memorable, effective and enjoyable interaction. As Nielsen refers these characteristics influence systems' acceptability by users (Nielsen, 1993). Figure 1 schematically represents the relationship of these particular characteristics with others that influence system usability.

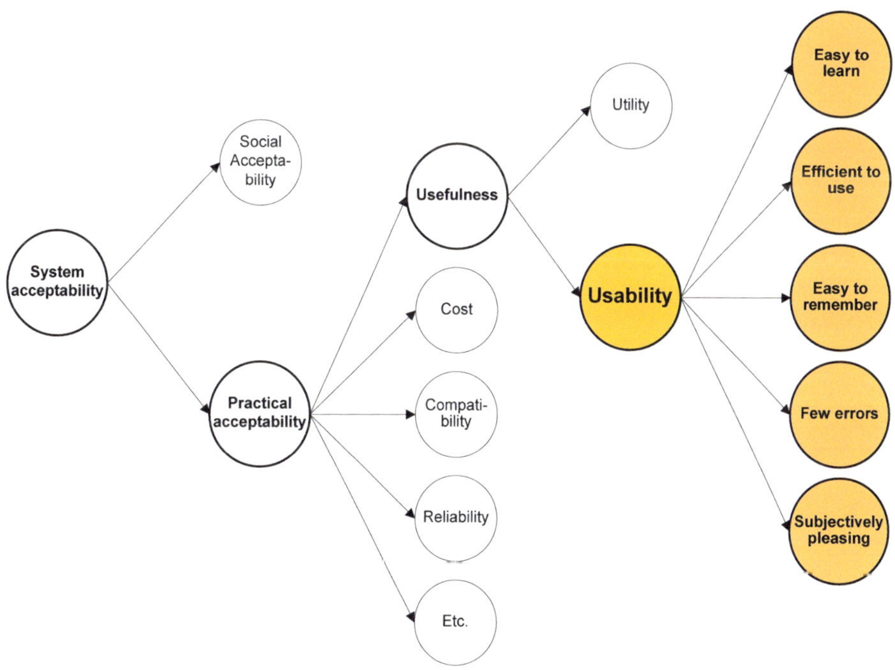

Fig. 1. A Model of the Attributes of System Acceptability (Nielsen, 1993).

Therefore, because of their influence in the usability of a system, it is important to define the concepts inherent to this set of characteristics (Nielsen, 1993):

- Ease to learn - the system must be intuitive, i.e. easy to use, allowing even an inexperienced user to be able to work with it satisfactorily;
- Efficiency of use - the system must have an efficient performance, allowing high productivity, i.e., the resources spent to achieve the goals with accuracy and completeness should be minimal;
- Memorability - the use of the system must be easy to remember, even after a period of interregnum;
- Errors frequency - the accuracy and completeness with which users achieve specific objectives. It is a measure of usage, i.e. how well a user can perform his task (e.g. set of actions, physical or cognitive skills necessary to achieve an objective);
- Satisfaction - the attitude of the user towards the system (i.e., desirably a positive attitude and lack of discomfort). Ultimately measures the degree to which each user enjoys interacting with the system.

According to Jordan (1998), when designing a product to achieve an appropriate usability developers should adopted the following 10 principles:

1. Consistency - similar tasks are performed in the same way;
2. Compatibility - the method of operation is compatible with the expectations of users, based on their knowledge of other types of products and the "outside world";
3. Consideration of user resources - the operation method takes into account the demands imposed to the resources of users during the interaction;
4. Feedback - actions taken by the user are recognized and a meaningful indication of the results of such activities is given;
5. Error Prevention and Recovery - designing a product so that the user likely to err is minimized and that, if errors occur, there may be a quick and easy recovery;
6. User Control - user control over the actions performed by the product and the state in which the product is are maximized;
7. Visual Clarity - the information displayed can be read quickly and easily without causing confusion;
8. Prioritization of Functionality and Information - the most important functionality and information are easily accessible by users;
9. Appropriate Transfer of Technology - appropriate use of technology developed elsewhere in order to improve the usability of the product;
10. Explicitness - offer tips on product functionality and operation method.

The design has also to consider the finite capability of humans to process information, to take decisions, and to act accordingly. These human characteristics have been thoroughly studied in the last decades, considering the Human Computer Interaction. Researchers that became a reference are, for instance, Hick (1952), Fitts (1954), or Miller (1956).

William Hick was a pioneer of experimental psychology and ergonomics. One of his most notorious researches was focused on the time a person takes to make a decision as a result of the possible alternatives, considering the cognitive information capacity, which was expressed as formula known as the Hick's Law (Hick, 1952).

Paul Fitts was a psychologist and a pioneer in human factors, which developed a mathematical model of human motion, known as Fitt's Law, based on rapid aimed movements (Fitts, 1954). This model is used, in the realm of ergonomics and human-computer interaction, to predict the time required to rapidly move to a target area, for instance to point with a hand or a finger, or with a pointing device in a computer interface.

George Miller was a cognitive psychologist that studied the average capacity of the human working memory to hold information. His studies concluded the number of objects an average person can hold is 7 ± 2 (Miller, 1956). This is known as the Miller's Law or the "magical number 7". One relevant consequence of this finding relates with the ability of humans to evaluate and judge alternatives, which is limited to 4 to 8 alternatives.

Accommodating all these research contributions in a set simple of design principles is problematic; therefore a different approach is the definition of heuristics for the assessment of the interfaces usability. An example of such approach is the work of (Gerhardt-Powals, 1996) that developed a set of heuristics to improve performance in the use of computers, which includes the following rules:

- Automate unwanted load:
  - Free cognitive resources for high-level tasks;
  - Eliminate mental calculations, estimations, comparisons, and unnecessary thinking.
- Reduce uncertainty:
  - Display data in a clear and obvious format.
- Condense the data:
  - Reduce the cognitive load, low-level aggregated data turning them into high-level information.
- Present new information with meaningful ways to support their interpretation:
  - Use a familiar framework, making it easier to absorb;
  - Use day-to-day terms, metaphors, etc..
- Use names that are conceptually related to functions:
  - Context-dependent;
  - Trying to improve recall and recognition;
  - Grouping data consistently significantly reduces the search time.
- Limit data-oriented tasks:
  - Reduce time spent in acquiring raw data.
  - Make the appropriate use of colour and graphics.
- Include only information on the screens that the user needs at any given time.
- Provide multiple coding of data, where appropriate.
- Practice a judicious redundancy.

A software program developed taking into account usability principles offers advantages, as decreased time to perform a task; reduced number of errors; reduced learning time, and improved satisfaction of system's users.

3. Reference standards on Usability

The international standard reference on the Usability is ISO 9241 - Part 11 from the International Organization for Standardization (ISO 9241, 1998). ISO 9241-11 emphasizes the usability of computers is dependent on the context of use, i.e., that the level of usability achieved depends on the specific circumstances in which the product is used. The context of use includes users, tasks, equipment (hardware, software and materials) and the physical and social environment, since all these factors can influence the usability of a product within a working system. Measures of performance and user satisfaction are used to evaluate the work system as a whole, and when the focus of interest is a product, these measures provide information about the usability of the product in the particular context of use provided by the work system. The performance and user satisfaction can also be used to measure the effect of changes in other components of the work system. Figure 3 shows schematically the set of factors to consider in evaluating the usability of a system, within the framework of ISO 9241-11.

ISO/IEC FDIS 9126-1 (ISO/IEC9126-1, 2000) follows the same general line. This standard for software quality that suggests a model based on quality attributes, divided into six main features, and the usability of them. According to this standard, usability is "the capability of the software product to be understood, learned, used and attractive to the user, when used under specified conditions".

This definition reinforces the idea that a product has no intrinsic usability, only a capability to be used under specified conditions (in a particular context). Usability depends on who are the users, what are their goals and where the users perform their tasks. Therefore, to analyze the usability of a software product, it is necessary to identify who are the users and what are its characteristics; what are the needs of users and what are their tasks; and what is the environmental context (social, organizational and physical).

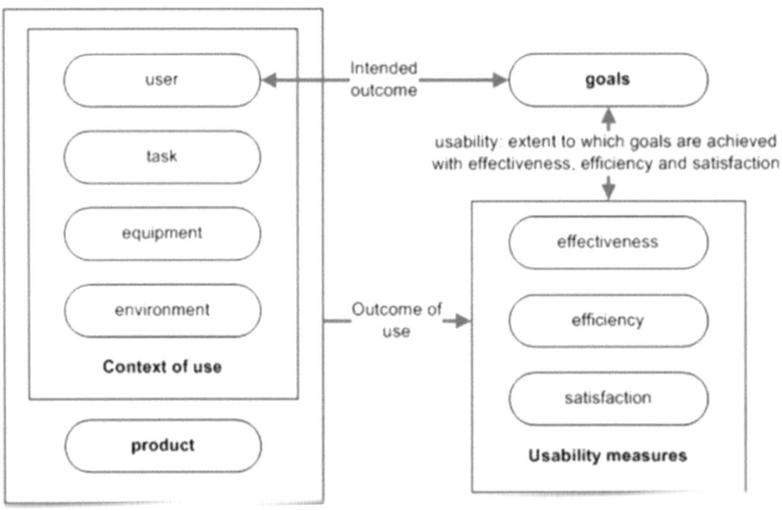

Fig. 2. Usability framework, according to ISO 9241-11 (ISO 9241, 1998).

Generally, the usability is evaluated based on the following dimensions (ISO 9241, 1998):

- Effectiveness (i.e., accuracy and completeness with which users achieve specified goals) as measured by success/failure that presents a user in the use of a product (e.g.,% of tasks completed, error rate or ratio hits /failures);

- Efficiency (i.e., resources expended in relation to the accuracy and completeness with which users achieve goals) as, for example, the time to complete the task, workload (physical and mental), deviations from the critical path or error rate;
- Satisfaction (i.e., freedom from discomfort and positive attitudes while using the product), as based on subjective judgments, e.g. ease of use (absolute or relative), usefulness of features, or like/dislike the product.

4. User-centred design

One approach to the use of the concept of software usability is the user-centred design. The user-centred design is a structured development methodology that focuses on the needs and characteristics of users, should be applied from the beginning of the development process in order to produce applications software more useful and easier to use (Averboukh, 2001); (Nunes, 2006).

According to ISO 13407 (ISO 13407, 1999), there are four key activities related to user-centred design, which should be planned and implemented in order to incorporate the requirements of usability in the process of software development (see Figure 3). The activities aim to:

- Understand and specify context of use;
- Specify the user and organizational requirements;
- Produce design solutions;
- Evaluate design against requirements.

These activities are performed iteratively, with the cycle being repeated until the usability goals have been achieved.

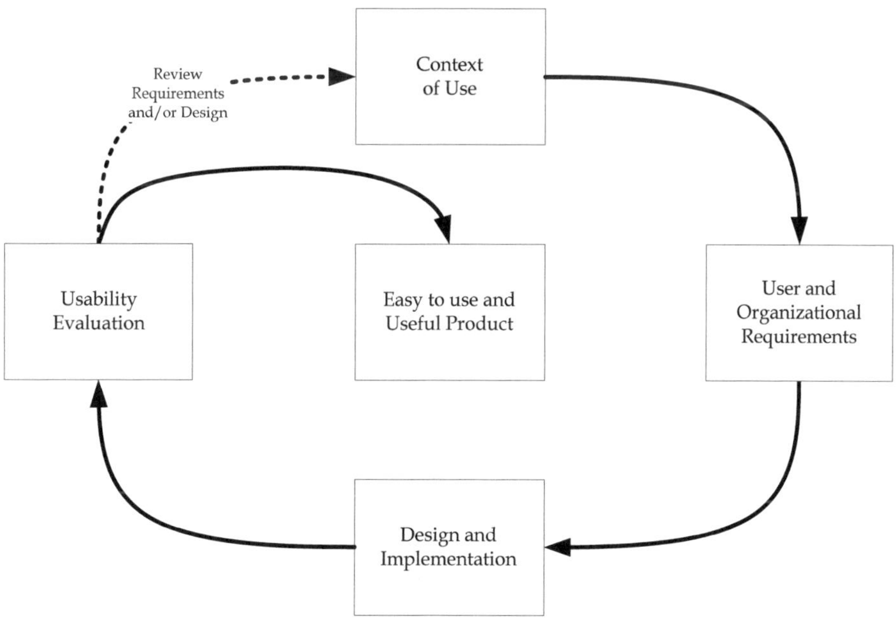

Fig. 3. Activities of user-centred design, adapted from ISO 13407 (ISO 13407, 1999).

According to (Howarth et al., 2009), the Usability Engineering process, which aims to implement the activities mentioned above regarding usability evaluation, includes (Figure 4):

- Identify and record critical usability data;
- Data analysis;
- Preparing the report of the evaluation results.

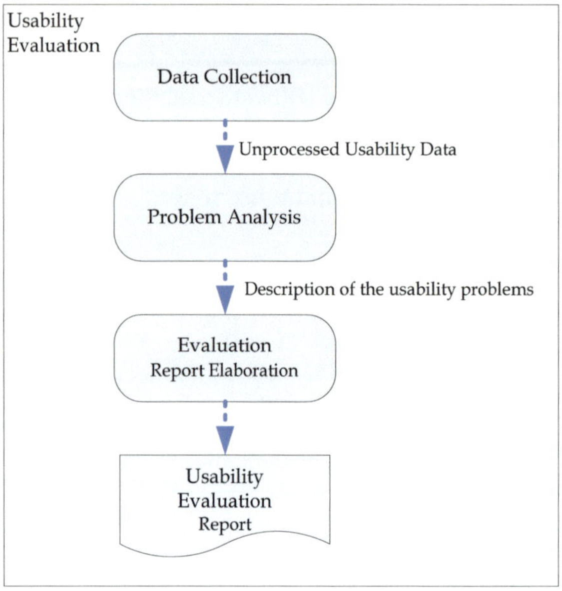

Fig. 4. Usability Evaluation, adapted from (Howarth et al., 2009).

In the section below that discusses guidelines and best practices are some recommendations on testing procedures, and reporting on the description of usability problems.

5. Methodologies and tools to support user-centred design

Usability analysis can occur at various stages of the development of a product (i.e., design, development and after implementation), although, hopefully, this analysis must be present at all stages, and should be iterative, allowing a continuous evolution of product quality.

There are several possible approaches to evaluate the usability, based, for example, on observation of users, application of questionnaires to users or analytical methods. The observation can be made in laboratory, but since the context of use is very important in usability studies, performing the study in the working environment where the system is intended to be used should be considered.

The assessment should draw on a representative sample of users for whom the system was designed.

Table 1 shows and describes a set of methodologies and tools for evaluating the usability considering, namely the phase(s) of development to which they apply, the amount of resources they require and how they are applied (based on (Usability Partners, 2011)). In this source, which offers a more complete list of tools or methods (in a total of 38 alternatives) it is possible to notice that most of the tools/methods are dedicated to a specific phase of project development (8 for the design phase, 10 for the preliminary design and prototyping, and 8 for the test phase and evaluation), 11 methods can be applied in two different phases, and only one (group discussion) is suitable to be applied in three phases.

Considering the early design and prototyping phase, the introduction of software development packages containing strong tools for developing the user interfaces, made easier and faster the prototyping of the graphical user interface (GUI) and of the basic interaction functionalities, turning almost obsolete other prototyping methods such as the paper- or video-based prototyping. Naturally, having real GUI prototypes helps the task of evaluating the usability of the products.

It should also be considered that there are many commercial support tools available to aid Usability Engineering activities. In most cases they are platforms for processing observational records. Some examples are:

- Morae, from TechSmith Corporation's (http://www.techsmith.com/morae.asp);
- Logger Egg, from Egg Studios LLC (http://www.ovostudios.com/ovologger.asp);
- Spectator, from BIOBSERVE GmbH (http://www.biobserve.com/products/spectator/index.html);
- Remote Usability Tester, from Pidoco (https://pidoco.com/en/benefits/products/remote-usability-tester).

Tool/Method	Stage in development			Resources required (Low, Med., High)	Purpose/ Strength	Summary Description
	Context & user requirements	Early design & prototyping	Test and evaluation			
Brainstorming		X		L	Produce design ideas	This is a group creativity technique (Osborn, 1953) by which a group of experts meet seeking to spontaneously generate new ideas by freeing the mind to accept any idea that is suggested. At the end a set of good ideas is generated.
Cognitive workload			X	L	Assesses if mental effort is acceptable	Cognitive workload (mental effort) can be measured subjectively using questionnaires which ask users how difficult they find performing a specific task. Examples of questionnaires are Subjective Mental Effort Questionnaire (Zijlstra, 1993) and the Task Load Index (NASA, 1986).
Cognitive walkthrough		X	X	M	Checks structure and flow against user goals	Cognitive walkthrough (Wharton et al., 1994) is a usability inspection method whose objective is to identify usability problems, focusing on how easy it is for new users to accomplish pre designed tasks. The reactions/comments of the users as the walkthrough proceeds are recorded.

Table 1. Examples of methods & tools for user-centered design, adapted from (Usability Partners, 2011).

Tool/Method	Stage in development: Context & user requirements	Early design & prototyping	Test and evaluation	Resources required (Low, Med., High)	Purpose/ Strength	Summary Description
Context of use analysis	X			L	Specifies user, tasks and environment characteristics	Context of use analysis (Thomas & Bevan, 1996) is a technique used for eliciting detailed information on user, tasks and environment. This information is collected during meetings of product stakeholders, which should occur early in the product lifecycle. The results should being continually updated and used for reference. Questionnaires can be used to evaluate current systems as an input or baseline for development of new systems.
Eye-tracking			X	H	Analyzes how users look at parts of an interface	Eye-tracking is a procedure for measuring either the point where we are looking or the motion of an eye relative to the head, using an eye tracker (Nielsen & Pernice, 2009). This method can be used to detect what users look at when interacting with an interface.
Heuristic evaluation		X	X	L	Provides feedback on user interfaces	Heuristic evaluation is a usability inspection method that helps to identify usability problems. It involves users, at least three, looking at an interface and judging its compliance with recognized usability principles/guidelines (the "heuristics"). The most well know heuristics are Nielsen Heuristics (Nielsen, 1994). Often the users are asked to rank the identified usability problems in terms of severity.
Task analysis	X			M	Analyses current user work in depth	Defines what a user is required to do (actions and/or cognitive processes) to achieve a task. A detailed task analysis can be conducted to understand a system and the information flow within it. Failure in performing this activity increases the potential for costly problems in the develop-ment phase. Once the tasks are defined, the functionality required to support the tasks can be accurately specified.
SUMI - Software Usability Measurement Inventory			X	L	Provides an objective way of assessing user satisfaction with software	SUMI is a method of measuring software quality from the end user's point of view (Kirakowski, 1994). Is based on a psychometric questionnaire (with 50 statements) designed to collect subjective feedback from users. The SUMI data is analysed by a program called SUMISCO. The raw question data is coded, combined, and transformed into a Global subscale, and five additional subscales: Efficiency, Affect, Helpfulness, Controllability, and Learnability. SUMI is mentioned in the ISO 9241 as a recognised method of testing user satisfaction. It is translated into several languages, for instance to Portuguese (Nunes & Kirakowski, 2010).
WAMMI - Web site Analysis and Measurement Inventory			X	L	Provides an objective way of assessing satisfaction w/ a web site	WAMMI measures user-satisfaction by asking website users to compare their expectations with what they actually experience on the website. It is based on standarised 20-statement questionnaire. This method uses five key scales: Attractiveness, Controllability, Efficiency, Helpfulness and Learnability and an overall Global Usability Score for how well visitors rate the website (http://www.wammi.com/ index.html).

Table 1. (cont.). Examples of methods & tools for user-centered design, adapted from (Usability Partners, 2011).

Nowadays new types of interface technology and forms of interaction are gaining importance, for instance touch and multitouch screens and gestures interaction devices. The use of touch screens has several potential benefits, usually because they are intuitive, easy to use and flexible, reducing the need of other input devices (e.g., keyboards, mouse). Touch screens are particularly adequate for devices that require high mobility and low data entry and precision of operation. This is typically the case of tablets and smartphones. Other examples of applications where touch screens are gaining terrain are information kiosks or checkout terminal.

However, designing for touchscreens presents some usability challenges. For instance, designers must take into account issues such as: fingers/hand/arm can hide the screen, the lack of tactile feedback, the parallax error resulting from the angle of view or the display may be overshadowed by dirt, stains or damage on the screen or on the protective film.

To the best of our knowledge, currently there is no usability assessment methodologies specifically developed to this type of interfaces. This fact has not prevented the usability studies multitouch devices, such as the studies by Budiu and Nielsen (Budiu & Nielsen, 2011) on the usability of applications iPad. These studies were based on methods commonly applied to other types of screens. Also (Heo et al., 2009) analyzing the question of usability of mobile phones covers a range of issues that are relevant also for other emerging interfaces such as touch and multitouch screens.

Guidelines \ Reference	(Haywood & Reynolds, 2008)	(HHS, 2006)	(Lea, 2011)	(Microsoft, 2009)	(MSDN, 2011)	(Sjöberg, 2006)	(Waloszek, 2000)
Touch	X		X	X	X	X	X
Multitouch			X	X	X		
Multi-users				X			
Mobile Devices	X						
Web Design		X					
Controls Usage	X	X		X	X	X	X
Controls Dimensions	X	X	X		X	X	X
Controls layout and spacing		X		X	X	X	
Interaction			X	X	X	X	X
Touch gestures			X	X	X		X
Error Tolerance					X		X
Screen layout	X	X		X		X	X
Feedback			X	X		X	X
Biomechanical considerations			X			X	
Design process & Usability Testing		X					

Table 2. Summary of references containing Guidelines applicable to touch and multitouch devices.

6. Guidelines and good practices

As previously mentioned, the use of touch devices, and particularly multitouch is recent and not yet widely adopted by most users. Despite the touch devices are common in kiosks or cash registers, the set of applications used and the requirements that have to obey is reduced. Therefore, the body of references that discuss generic usability for multitouch devices and present guidelines and best practices for the design of applications is still not significant. Despite this limitation, some references applicable to touch and multitouch devices that offer recommendations on basic features for these interfaces is presented in Table 2. The type of references differs significantly, as well as their emphasis on different types of interfaces. For instance, Microsoft (2009) and MSDN (2011) focus on touch applications. (HHS, 2006) is not specifically dedicated to touch interfaces, is a compilation of about 500 general guidelines to consider in developing applications, including the ones devoted to Web environments.

In addition to these references, others such as (Largillier et al., 2010) and (Meador et al., 2010), discuss the evaluation of characteristics of multitouch devices but are not exclusively focused on the guidelines perspective.

Tables 3 and 4 offer other elements that might be relevant to specific aspect of usability evaluation, namely related with usability testing and usability problem reporting as suggested by (Kreitzberg & Little, 2009) and (Capra, 2006).

About the Tests	About Reporting Process
1. Decide what to test & develop test objectives; 2. Decide the type of participant in the tests and how many people to recruit; 3. Decide on the tasks and to use an experimental design and produce a script; 4. Decide if video recordings are made. If so, consider the need for a consent form; 5. Consider a questionnaire pre- and post-test and an introduction and debriefing interview; 6. Recruit & schedule participants' involvement; 7. Test the script, and materials to make sure that the testing process will run smoothly; 8. Perform the Testing; 9. Analyze results and prepare the report and recommendations.	Reports on Usability testing should cover: 1. The objectives of the test and an executive summary; 2. As participants were recruited; 3. Description of the tasks of the project and the test facility used; 4. The main problems found and recommendations to address them; 5. Small problems discovered and recommendations to address them; 6. Direct quotations of participants; 7. Recommendations and next steps.

Table 3. Recommendations for testing Usability, adapted from (Kreitzberg & Little, 2009).

• Be clear and precise, avoiding long words and jargon:	• Define the terms used, and be concrete, not vague; • Be practical and not theoretical; • Use descriptions of what people do not like to experts in HCI; • In the description avoid as details that no one wants to read; • Describe the impact and severity of the problem; • Describe how the task affects the user; • Describe how often the problem occurs, and the components that are affected.
• Base the findings on data	• State how many users experienced the problem and how often; • Include objective data, both quantitative and qualitative, such as the number of times a task was attempted or the time spent on task; • Provide traceability of the problem in the observed data.
• Describe the cause of the problem	• Describe the main usability issue involved in the problem; • Avoid assumptions about the cause of the problem or the thoughts of the user.
• Describe the actions the user observed	• Include background information about the user and the task; • Include examples, such as user navigation flow through the system, user's subjective reactions, screenshots, and success / failure in performing the tasks; • State whether the problem was user reported or experimenter observed.
• Describe a solution to the problem	• Offer alternatives and tradeoffs; • Be specific enough so as to help but without imposing a solution; • Complementary to the principles of design for usability.

Table 4. Guidelines to describe Usability problems, adapted from (Capra, 2006).

7. Conclusions

Usability is a critical aspect to consider in the development cycle of software applications, and for this purpose, user-centred design and usability testing must be conducted. The design and testing cannot ignore the context of use of software, whose knowledge is essential.

Usability of a system is characterized by its intuitiveness, efficiency, effectiveness, memorization and satisfaction. Good usability allows decreasing the time to perform tasks, reducing errors, reducing learning time and improving system users' satisfaction.

Usability, process design and development of software have necessarily to be framed by the characteristics of users, tasks to perform and environmental context (social, organizational and physical) for which the product is intended to.

The development of a product must consider the 10 basic usability principles: consistency, compatibility, consideration by the resources of the user, feedback, error prevention and error recovery, user control, clarity of vision, prioritization of functionality and information, appropriate technology transfer, and clarity.

There is a wide range of tools and methodologies for identifying and evaluating the usability of a system, thus contributing directly or indirectly, for its improvement. The selection of these tools and methodologies depends on the objective to achieve, which usually is related to the development phase the system is in. Some approaches are better suited to the design stage (e.g., analysis of context of use and task analysis), while others are more suited to early stages of development and prototyping (e.g., brainstorming, prototyping) and others for the evaluation and testing (e.g., analytical and heuristic evaluations, SUMI).

Finally, in developing a solution one has to consider the guidelines and best practices that are relevant, taking into account the specific context. There is a vast literature on generic guidelines for usability. As mentioned, the body of reference for touch and multi-touch interfaces is very limited, since this is a quite new type of user interface. Nevertheless there is a significant number of sources and formal or industrial standards that may be adopted.

8. Acknowledgments

The work described in this paper was developed under the project BrainMap, a partnership between Viatecla, University of Évora and Centre of Technologies and Systems of Uninova, supported by the QREN - Programa Operacional de Lisboa.

9. References

Averboukh, E. A. (2001). Quality of Life and Usability Engineering. In: *International Encyclopedia of Ergonomics and Human Factors*, W. Karwowski (Ed.), pp. 1317-1321, Taylor & Francis

Budiu, R., Nielsen, J. (2011). Usability of iPad Apps and Websites. First Research Findings. Available at http://www.nngroup.com/reports/mobile/ipad/.Nielsen Norman Group

Capra, M. (2006). Usability Problem Description and the Evaluator Effect in Usability Testing. Unpublished Dissertation, Virginia Tech, Blacksburg, VA

Dillon, A. (2001). Evaluation of Software Usability. In: *International Encyclopedia of Ergonomics and Human Factors*, W. Karwowski (Ed), pp. 1110-1112, Taylor & Francis

Fitts, P. M. (1954). The information capacity of the human motor system in controlling the amplitude of movement. *Journal of Experimental Psychology*, Vol. 47, No 6, pp. 381-391

Gerhardt-Powals, J. (1996). Cognitive engineering principles for enhancing human - computer performance. *International Journal of Human-Computer Interaction*, Vol. 8, No 2, pp. 189–211

Gould, J. D. (1995). How to Design Usable Systems (Excerpt). In: *Readings in Human-Computer Interaction: Toward the Year 2000*, R. M. Baecker, J. Grudin, W. A. S. Buxton and S. Greenberg (Ed), pp. 93-121, Morgan Kaufmann Publishers, Inc.

Haywood, A., Reynolds, R. (2008). Usability guidelines. Touchscreens. Available at http://experiencelab.typepad.com/files/design-guidelines-touchscreens-1.pdf.Serco, ExperienceLab

Heo, J., Ham, D.-H., Park, S., Song, C., Yoon, W. C. (2009). A framework for evaluating the usability of mobile phones based on multi-level, hierarchical model of usability factors. *Interacting with Computers*, Vol. 21, pp. 263–275

HHS (2006). Research-Based Web Design & Usability Guidelines. Available at: http://usability.gov/guidelines/guidelines_book.pdf. .U.S. Department of Health and Human Services' (HHS)

Hick, W. E. (1952). On the rate of gain of information. *Quarterly Journal of Experimental Psychology*, Vol. 4, pp. 11-26

Howarth, J., Smith-Jackson, T. , Hartson, R. (2009). Supporting novice usability practitioners with usability engineering tools. *Int. J. Human-Computer Studies*, Vol. 67, No 6, pp. 533-549

ISO 9241. (1998). *Ergonomic requirements for office work with visual display terminals (VDTs) - Part 11: Guidance on Usability*. International Organization for Standardization

ISO 13407. (1999). *Human-centred design processes for interactive systems*. International Organization for Standardization

ISO/IEC9126-1. (2000). *ISO/IEC TR 9126- 1. Software engineering - Product quality - Part 1: Quality model*. International Organization for Standardization

Jordan, P. (1998). *An Introduction to Usability*, Taylor & Francis

Kirakowski, J. (1994). The Use of Questionnaire Methods for Usability Assessment.

Kreitzberg, C. B. , Little, A. (2009). Usability in Practice. Usability Testing. Available at http://msdn.microsoft.com/en-us/magazine/dd920305.aspx. *MSDN Magazine*, Jul

Largillier, G., Joguet, P., Recoquillon, C., Auriel, P., Balley, A., Meador, J., Olivier, J., Chastan, G. (2010). Specifying and Characterizing Tactile Performances for Multitouch Panels: Toward a User-Centric Metrology. Available at http://www.leavcom.com/stantum_092310.php. Leavitt Communications.

Lea, J. (2011). Unity Gesture UI Guidelines. Available at: https://docs.google.com/View?id=dfkkjjcj_1482g457bcc7.

Meador, J., Auriel, P., Chastan, G. (2010). How to Evaluate Multitouch While Standing in a Store. Available at http://www.stantum.com/medias/whitepapers/How_to_Evaluate_Multi-Touch_While_Standing_in_a_Store.pdf. Stantum.

Microsoft. (2009). User Experience Guidelines. User Interaction and Design Guidelines for Creating Microsoft Surface Applications. Available at http://download.microsoft.com/download/A/8/E/A8ED7036-10DC-48AB-9FE8-D031D9E28BBB/Microsoft Surface User Experience Guidelines.pdf

Miller, G. A. (1956). The magic number seven, plus or minus seven. *Psychological Review,* Vol. 63, No 2, pp. 81-97

MSDN (2011). Touch. Available at http://msdn.microsoft.com/en-us/library/cc872774.aspx. Microsoft

NASA (1986). *Collecting NASA Workload Ratings: A Paper-and-Pencil Package. NASA-Ames Research Center, Human Performance Group* Moffet Field, CA, NASA-Ames Research Center, Moffet Field, CA.

Nielsen, J. (1993). *Usability Engineering,* Academic Press,

Nielsen, J. (1994). Heuristic evaluation. In: *Usability Inspection Methods,* J. Nielsen and R. L. Mack (Ed), John Wiley & SonsNew York

Nielsen, J., Pernice, K. (2009). *Eyetracking Web Usability,* New Riders Press ISBN-10: 0-321-49836-4

Nunes, I. L. (2006). Ergonomics & Usability - key factors in knowledge society. *Enterprise and Work Innovation Studies,* Vol. 2, pp. 87-94

Nunes, I. L., Kirakowski, J. (2010). Usabilidade de interfaces – versão Portuguesa do Software Usability Measurement Inventory (SUMI) [Interfaces Usability – Portuguese version of the Software Usability Measurement Inventory (SUMI)]. Proceedings of Occupational Safety and Hygiene (SHO10), 978-972-99504-6-9, Guimarães - Portugal

Osborn, A. (1953). *Applied Imagination: Principles and Procedures of Creative Problem Solving.* New York, Charles Scribner's Sons, ISBN 978-0-02-389520-3, New York

Sjöberg, S. (2006). A Touch Screen Interface for Point-Of-Sale Applications in Retail Stores. Available at: http://www8.cs.umu.se/education/ examina/Rapporter/ SamuelSjoberg.pdf.MSc Thesis. Umea University, Sweden.

Thomas, C. & Bevan, N. (1996). *Usability Context Analysis: A practical guide,* Serco Usability Services

Usability Partners. (2011). Methods. Selecting tools and methods. Available at: http://www.usabilitypartners.se/about-usability/methods.

Waloszek, G. (2000). Interaction Design Guide for Touchscreen Applications. Available at: http://www.sapdesignguild.org/resources/tsdesigngl/TSDesignGL.pdf.SAP Design Guild Website

Wharton, C., Rieman, J., Lewis, C., Polson, P. (1994). The cognitive walkthrough method: a practitioner's guide. In: *Usability Inspection Methods,* J. Nielsen and R. L. Mack (Ed), pp. 105-140, John Wiley & Sons, ISBN 0-471-01877-5

Zijlstra, F. R. H. (1993). *Efficiency in Work Behaviour: a Design Approach for Modern Tools.* Delft, Delft University Press, Delft

8

User Experience Design: Beyond User Interface Design and Usability

Wei Xu*
Intel Corporation
USA

1. Introduction

This chapter first discusses major challenges faced by current user-centered design (UCD) practices. A user experience design (UXD) framework is then proposed to address these challenges, and three case studies are analyzed to illustrate the UXD approach and formalize the UXD processes. Finally, this chapter discusses future research needs.

2. Major challenges faced by UCD

User-centered design has been widely practiced by user experience (UX) professionals for many years, and a variety of methods have been used to facilitate UCD practices (Nielsen, 1993; see Chapter X of this book). *UX professionals* herein refers to the people who are practicing UCD, including human factors engineers, UX designers, human-computer interaction (HCI) specialists, usability specialists, and the like. The philosophy of UCD places emphasis on the end users when developing usable products (e.g., applications, services). It focuses on users by understanding the users, learning their environments and contexts of their usages, and realizing their needs in usable products.

Much progress has been made toward improving UCD practices and in increasing UCD influences on product development since its inception (Xu, 2001, 2003, 2005, 2007; Xu, Dainoff, & Mark, 1999). For instance, UX professionals now are involved in product development earlier than before; they contribute to definitions of product requirements, instead of just running ad-hoc usability testing of the user interface (UI) design; and they drive the design usability work by defining and tracking usability success metrics.

Overall, current UCD practices aim primarily at the usability of the product UI to achieve usability goals, such as ease of use, efficiency, reduced error, easy to remember, and user satisfaction (Nielsen, 1993). They identify user needs, conduct task analysis, define UI concepts, and conduct interactive prototyping and usability testing to optimize the UI design. The focus on UI design and usability has demonstrated UCD's contributions to the traditional approach to product development that focuses on system and product functionality. However, current UCD practices still prove challenging, which limits the

* wei.xu@intel.com

potential to make greater contributions to product development. Major challenges are discussed below.

2.1 Challenge 1: Not effectively addressing total user experience

Norman (1999) coined the classic definition for *UX*: "all aspects of the user's interaction with the product: how it is perceived, learned, and used." Clearly, this definition suggests that UX is beyond UI design and usability. Norman's definition of UX is extended herein to a scope of total user experience (TUX) in a broader UX ecosystem context, as illustrated in Figure 1.

First, a clear understanding of the UX ecosystem clarifies the definition of UX in such a context. A UX ecosystem as defined herein is twofold. First, end users receive their UX from a product throughout a UX lifecycle across various stages, such as early product marketing (how it is perceived), use of the product (how easy it is to use), training and user help (how it is learned), support (how the user is supported), upgrade (how the user gets new versions), and so on.

Second, users receive their UX through all aspects (touch points) of their interactions with a product across all UX lifecycle stages, including anything related to the product, such as functionality, workflow, UI design and usability, online help, user manual, training, user support, service content, and the like. Multiple touch points may coexist in a single UX lifecycle stage of using a product. For instance, in the UX lifecycle stage, users may experience the product's functionality, UI design and usability, online help, and so on.

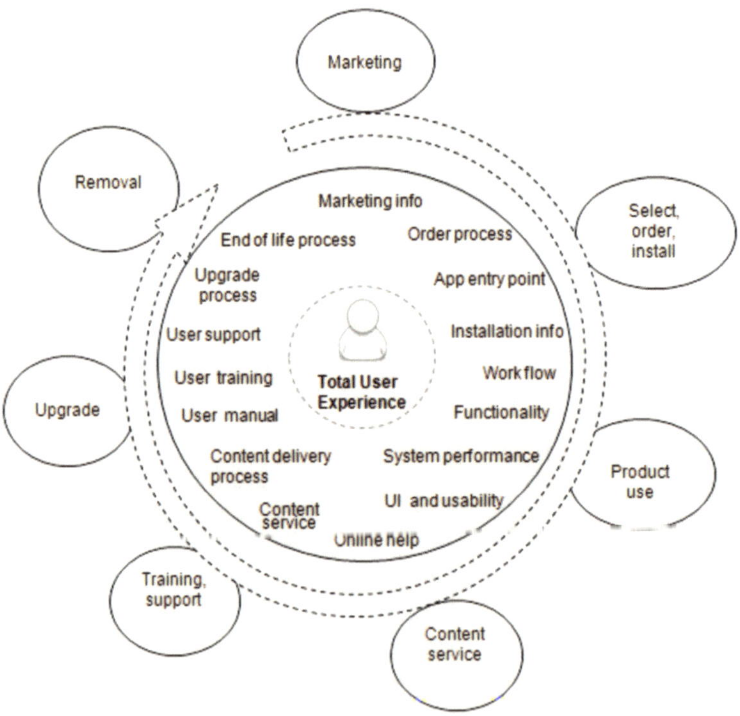

Fig. 1. The total user experience (TUX) concept in a UX ecosystem context.

User-experience ecosystems vary in terms of scale and nature across product domains. For instance, Apple Inc. has been building a macro UX ecosystem across a variety of product lines. In addition to its easy-to-use UI for individual products, Apple Inc. also focuses on TUX-related applications, content, service, and so forth. As part of TUX, its rich content (e.g., songs, apps) can be shared across different products, such as iPhone, iPad, and iPod, through its centralized iTunes service platform. In this case, the UX ecosystem spans from marketing and branding at the early stage, to purchasing, product use, post-buy support, updates, and then to the late stage of post-purchase content sharing across product lines. End users gain TUX not just from the UI of the individual products, but also from all touch points in such a macro UX ecosystem.

Also, individual products may have their own micro-UX ecosystems spanning from marketing and branding to post-sale services across all TUX touch points. For example, in a corporate IT setting, employees are often required to install enterprise applications. Employees may experience difficulties installing the application if multiple manual steps are required and the user manual is not usable, which may negatively affect employee productivity and result in support calls. Once installed, employees may experience ongoing difficulties without necessary self-help functionality, although the UI meets accepted design criteria. Thus, even if the application at some point allows employees to complete activities through its UI, other activities (such as the initial installation and ongoing support) can lead to an overall negative TUX.

Thus, from the perspective of a UX ecosystem, end users actually receive UX through an overall TUX, instead of through any single interaction touch point with a product. This implies that UX is a continuous involvement through various interaction touch points with a product across its UX lifecycle; any breakdown of these touch points would negatively impact TUX and cause a failure in delivering an optimal UX.

Obviously, UI design and usability are only one of the key interaction touch points within TUX. From a marketing competition perspective, the success of a product in today's market no longer depends only on the UI design and usability; it actually depends on how well TUX is delivered to end users within its UX ecosystem. Apple Inc.'s success in the market is a good example. If we primarily focused on UI design and usability in the practices, we would not be able to deliver optimal TUX to end users.

In addition, there are no systematic approaches and methods that are available to guide current UCD practices to address TUX from a broad UX ecosystem perspective. There is also a lack of organizational culture that can effectively facilitate collaborations among various TUX stakeholders to address TUX. TUX stakeholders include various owners of these TUX touch points, such as professionals in UX, marketing, training, technology, and business. No single person in any of these individual areas could address all these TUX touch points, and optimal UX would not be achieved without joint efforts through collaborations among TUX stakeholders.

2.2 Challenge 2: Not predictively considering UX evolution to influence a product's strategic direction

A technology or business capability roadmap is a common method that matches short-term and long-term goals with specific technology or business solutions to help meet those goals.

These capability roadmaps largely determine the UX of a final product that delivers to end users. However, in current practices, the development of these roadmaps has been driven mainly by technology people (e.g., architects) and business people (e.g., product line managers). When developing the roadmaps, they focus on business and technology based on customer requirements and technological trends, but UX (e.g., user gaps, needs) is not fully considered. In most cases, the customer requirements basically come from business stakeholders and owners who may not be the real end users of the product to be built, so those customer requirements may not truly represent end users' requirements. Thus, there is a gap in the process from the UX perspective (Wooding & Xu, 2011).

From the perspective of the UX ecosystem, UX dynamically evolves in terms of user needs and usages. User needs and usages for a product advance over time in a sequential order, which may be influenced by improvements in technology and people's living conditions. One user need or usage may have to be satisfied before subsequent user needs and usages while the products' initial UXs are maturing; otherwise, optimal UX will not be delivered. Also, UX is predictable because those user needs and usages can be analyzed and defined based on data collected from end users. These sequential and predictable UX data may potentially help UX professionals influence technology and business capability roadmaps so that the capabilities of a product can be delivered in a sequential order to match the optimal UX sequence and, eventually, optimal UX can be delivered over time, as needed.

In current UCD practices, UX professionals focus on the end users of products, and they collect UX data, such as end user needs and usage models, through various UCD activities. However, the challenge for UX professionals, in most cases, falls into one of the following three scenarios: 1) They did not proactively conduct user research to fully understand user needs and usages, either short-term or long-term; 2) they did not leverage the collected UX data to generate predictive UX data in terms of user needs and usages over time; or 3) they defined the predictive UX data, but they either did not leverage the predictive UX data or did not have an opportunity to influence technology and business capability roadmaps at the early product planning stage.

Figure 2 illustrates the gap in developing technology and business capability roadmaps in current practices (see the left side of Figure 2). That is, without considering UX, a delivered

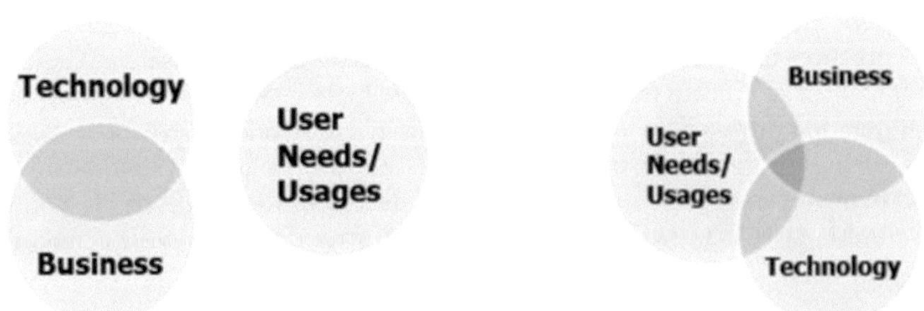

Gap in current practices

An optimal solution resides in the balanced overlapping area across the three areas

Fig. 2. A concept that demonstrates how the intersection of business, technology, and UX would impact a delivered solution.

product may provide great technical capabilities that match a predefined business strategy, but it may not be the product or capabilities that end users want and is therefore unusable. As a result, the product may fail to achieve the business strategy and goals, such as market shares and return on investment. In reality, there are many cases where conflicting requirements may occur among technology, business, and UX over a product's evolution process; an optimal solution always resides in a balanced overlapping across all three areas (see the right side of Figure 2). The size of the overlapping area varies depending on the scale of conflicts, and a best effort should be made to maximize this overlap.

Current UCD practices therefore do not predictively consider the evolutionary nature of UX over time in a context of its ecosystem, and they lose the opportunity to influence the development of technology and business capability roadmaps. A gap may have already existed from the very beginning, when technology and business people defined the strategic direction for current (at the time) and future products when developing their roadmaps. Lack of such an influence would limit UX professionals' work in a passive and tactical work mode only within the predefined scope of a current project. Such a work mode would not only limit UX professionals' ability to deliver the best UX in current release (because user needs may not be sequentially optimized), but may also limit UX professionals' long-term influences on the strategic directions of products.

2.3 Challenge 3: Not proactively exploring emerging UX to identify new UX landing zones

New components always emerge over time in any ecosystem. The UX ecosystem is no exception. A variety of new user needs and usages may emerge daily as their UXs mature. Although premature, some are emerging as patterns with valid usages that represent a new UX landing zone. Such a new UX landing zone, which may have been previously unknown, creates a potential marketing opportunity for a new product that meets user needs and usages. In today's competitive market, whoever captures a new valid UX landing zone early enough and builds a product at the right time may win the market. There are several cases of such success in today's market, such as certain types of tablets, netbook computers, and smartphones.

However, current practices in identifying market opportunities for new products are primarily driven by current market methods. These market methods are limited in terms of understanding actual UX and user behaviours in end users' real-life settings, because the data collections are based mainly on user opinions or preferences gathered through such methods as surveys and focus group sessions. These methods do not fully explore users' behaviours and usages in their real-life settings. In many cases, the things users say may not truly represent their needs.

On the other hand, in current UCD practices there are many methods available that help UX professionals identify actual user needs and usage models in a social-tech environment through direct user behavioural studies, such as ethnography and contextual inquiry. These identified user needs and usage models may lead to the creation of a new UX landing zone in the very early stage; that is, even before a product development lifecycle starts. However, although UX professionals have tried to get involved in the early stages of a product's lifecycle and have made great progress, UX professionals with current UCD practices are not proactive enough to explore emerging UX. Therefore, their contributions to the process of identifying new market opportunities are limited, where UX is not fully considered.

Once a new marketing opportunity is defined, platform architecture design begins as part of the product's requirements. The platform architecture determines the foundation for the technical capabilities (both hardware and software) of a product, which determines the human-computer interaction functionality and the UI technology that can be developed in order to design a usable product. For instance, computing platform architecture consists of a CPU (central processing unit), chipset, and system hardware and software, all of which determine the functionality and UI technology for an end product (e.g., laptop, tablet) that will be built based on that platform.

However, in today's practices, a technology-centric approach is typically used in defining platform architecture capabilities. In the case of defining the platform architecture for CPU, people used to focus on system performance (e.g., CPU computing speed) and did not pay enough attention to user needs to foresee the UI capabilities to be used in the end products that are built on the CPU, such as wireless, touch-screen UI, 3-D graphics, instant boost, and multimedia. Without these types of capabilities built into the platform architecture, original equipment manufacturers (OEM) (e.g., Dell, HP) cannot use the CPU to build these UI capabilities into their end products to meet user needs.

Although UX professionals (e.g., Dell or HP UX professionals) may participate early enough in the development of their own products by following their UCD process, lack of these types of fundamental platform architecture capabilities will restrict these UX professionals from developing rich UX for end users through their UI. Therefore, if there is a lack of UX considerations in defining platform architecture capabilities in the very beginning, delivered UX of an end product will be greatly impacted. Again, in current UCD practices, UX professionals typically are not involved at such an early stage.

In summary, UX professionals in current UCD practices are not proactive enough to explore new emerging UX in its ecosystem. Without UX professionals' involvement from the very beginning, a UX gap may already exist when people defined market opportunities for new products and platform architecture capabilities. In this case, UX professionals who work on the end product will not be able to deliver good UX to meet end user needs, no matter how much effort they put into following UCD, because the end product may have been wrongly defined without a valid UX landing zone in the first place, and/or the platform architecture may not provide necessary capabilities that support user interactions on the UI.

3. A user experience design (UXD) framework

To address these challenges faced by current UCD practices, a conceptual user-experience design (UXD) framework is proposed herein (see Figure 3). The UXD framework has its roots in user-centered design (UCD), but beyond UCD that primarily focuses on UI design and usability. As shown in Figure 3, the UXD framework expands its boundaries far beyond UCD; it approaches UX in the context of a broad UX ecosystem, including various UX components from emerging UX in the beginning, all TUX touch points across a product UX lifecycle, and future UX evolution. Specifically, the UXD framework characterizes the UXD approach as follows.

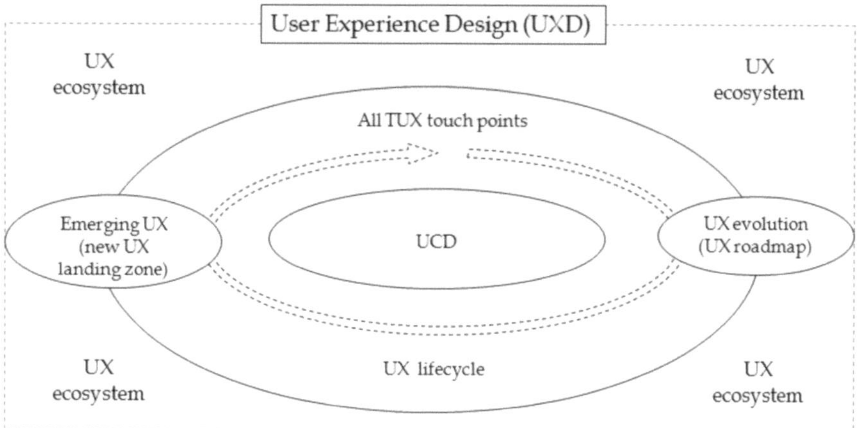

Fig. 3. The conceptual user experience design (UXD) framework in a UX ecosystem context.

- **It is a philosophy:** UXD addresses UX in the context of a broader UX ecosystem by emphasizing three aspects: (1) the emergence of new UX so that UX professionals may identify new UX landing zones in the very early stages of product development in order to influence market opportunities for new products and platform architecture definitions, instead of just executing UCD activities based on predefined products and predefined platform architecture with available UI technology in the UCD practices of the day; (2) the continuous nature of UX in terms of TUX; that is, continuous involvement through various TUX touch points with a product across all UX lifecycle stages, instead of focusing primarily on one single touch point at a single UX lifecycle stage, such as UI design and usability; and (3) the evolutionary nature of UX over time in terms of user needs and usages so that UX professionals can deliver predictable UX roadmaps to influence technology and business capability roadmaps in a long-term perspective, instead of a narrowed focus within the scope of a predefined current project for a near-term goal.
- **It is a process:** UXD leverages current UCD processes to deal with UX, such as a broader UX ecosystem. Beyond that, UXD requires additional processes to address all TUX touch points throughout a UX lifecycle, new emerging UX, and predictable UX. Overall, current UCD processes need to be enhanced to incorporate and facilitate some key UXD activities, such as UXD success scorecards and a tracking system. UXD requires much early involvement of UX professionals in the development lifecycle; it may ask UX professionals to execute activities beyond the scope of current individual projects. From a methodology perspective, UXD also continues to leverage current UCD methods with necessary enhancements. Besides, to be more user-centric and to support executions of the UXD approach, UXD requires the enhancement of conventional methods (e.g., training, marketing).
- **It requires great collaborations:** UXD requires great collaborations among UXD stakeholders who own each TUX touch point, such as people who own business processes, user training, user support desks, technology and business capability roadmaps, new marketing definitions, and platform architecture definitions. It is impossible for UX professionals to accomplish UXD goals without such collaborations. An organization culture should be established to facilitate such collaborations.

User-experience professionals at Intel and IBM, for example, have done some initial work that addressed the non–UI-related UX issues across UX lifecycle stages (Finstad et al., 2009; Swards, 2006; Vredenburg et al, 2001). The UXD framework presented here is intended to propose a formal framework in a more systematic way, and in a much broader context, from a UX ecosystem perspective. Besides, the label "UXD" has been used by others (e.g., Bowles & Bowles, 2010; Unger & Chandler, 2009). These versions of UXD vary slightly from one another, but all of these versions basically echo the current UCD approach that primarily focuses on UI design and usability. Therefore, these UXD approaches essentially do not differ from current UCD practices.

Thus, as compared to current UCD practices, the UXD approach intends to be: (1) more proactive by participating in much earlier stages of defining market opportunities and platform architecture; (2) broader through addressing the TUX in a UX ecosystem context; (3) more collaborative by partnering with various TUX owners; and (4) more predictive by developing UX roadmaps. Therefore, UX professionals can be more influential, creative, and strategic by practicing the UXD approach.

4. UXD practices and processes

This section discusses three case studies. Each one includes three parts. First, problem statements are described; second, details of the case study are discussed to illustrate how the problems have been addressed through a UXD solution in the practice; and finally, a UXD process is initially formalized.

4.1 Case Study 1: Effectively address total user experience

4.1.1 Problem statements

A few years ago, Intel planned to upgrade a large enterprise back-end database system. As a result, upgrades of some front-end, web-based applications were also required. The external vendor of the back-end system offered a front-end, web-based application suite at no cost. To save on costs during the economic downturn, as parts of the system upgrade program, the vendor's application suite was chosen to replace the existing web-based enterprise application suite (WEA 1). After WEA 1 was released, significant post-release issues were reported. Overall, end users perceived the upgrade as a step back, from a UX perspective. Two root causes were identified:

- **Vendor-side issues:** The application suite was the first-generation, web-based, front-end solution built by the vendor; the vendor had not done enough necessary UX work on it. As a result, the product was delivered with many UX issues across its UI design, business process, functionality, system integration, configuration capabilities, and user help materials, among others.
- **Enterprise-side issues:** As influenced by the overall cost policy used for the back-end system upgrade, a vanilla (i.e., no customization) approach was executed for the front-end application suite and UX work was not considered a high priority in the process.

To address the significant post-release issues, the phase 2 work (WEA2) kicked off. The human factors engineer (HFE) group was requested to provide support for WEA 2. A lead

HFE was assigned to the WEA 2 program. The HFE conducted a UX gap analysis based on WEA 1 post-release issues. The analysis clearly indicated that although there were many UI-related usability issues, UI usability issues only accounted for a small proportion (11%) of the total identified post-release issues. The overall identified issues were distributed across all aspects of TUX, including system and data integration, application functionality, business process, application configuration, system performance, online help, user support model, and marketing, among others. Obviously, if the project team just fixed all these UI usability issues, the team still would not be able to significantly enhance TUX.

4.1.2 The UXD solution

The HFE proposed a UXD solution for WEA 2, which was approved by program management. Three major steps were taken to facilitate the UXD process (Finstad et al., 2009).

- **Created a UXD team.** The HFE led the UXD team; members included representatives from such functional areas as across quality assurance, business process, transition change management (TCM), training, user support, and others. The HFE worked as facilitator of the team. Each of the UXD team members owned the planning and execution of the TUX component in his or her functional area.
- **Defined a TUX scorecard and a tracking process.** The TUX scorecard not only defined success criteria for usability as a typical UCD process (e.g., task completion, success rate), but also covered success criteria for other TUX aspects. Besides, various check points were defined across all these TUX touch points in alignment with the program lifecycle. Such a tracking process enabled the program management office (PMO) to closely monitor the progress of UXD and take necessary actions, if needed. This process also increased the overall awareness of a UXD culture within the program.
- **Included HFE as a member of the PMO.** This is different from conventional UCD practices, where UX professionals are typically embedded somewhere within a program as a project member. Becoming a PMO member helped promote UXD and increased the visibility of UXD work to the PMO.

Specifically, various UXD activities were executed as highlighted below:

- **Incorporated UXD into vendor selection:** During the selection of a new vendor for the application suite, UX requirements were incorporated into a vendor assessment scorecard and counted as 20% of the total score among the five components (i.e., business requirement fit, solution compliance, vendor viability, cost, and UX). A UX assessment template was defined to score various items across different TUX aspects, including UI usability, business process, training needs, online design, and others. The PMO made the final decision based on the total score among three candidate vendors. This ensured that UX was fully considered in the vendor selection process.
- **Leveraged UX data to optimize business processes:** The product from a new vendor was chosen, partially due to its flexible configuration capability of business processes as one of the advantages over others. In order to achieve the right balance between UX and business processes, four usability tests were conducted with various configured business processes. Eventually, an optimal business process was chosen based on a trade-off of decisions that achieved a streamlined business process with more intuitive

UI design, but without violating necessary business processes, such as legal requirements.

- **Collaborated with the vendor for major UI usability improvement:** Two critical usability issues were identified during the usability tests. The UXD team directly worked with the vendor and convinced them to fix the issues based on usability test data. This saved Intel substantial costs by avoiding customization coding work and helped other customers of the vendor.
- **Incorporated user-centric approach to conventional methods:** Influenced by the UXD approach, the training and user support teams shifted their focus from a conventional "quantity" approach (e.g., percentage of users trained) to a "quality" approach (e.g., effectiveness of the training delivered). The teams conducted training and support-need analysis across three target user segments and implemented effective training delivery methods based on the needs and priorities identified. Each training delivery (e.g., web-based training, in-classroom training) was tested through UX validations (e.g., surveys, usability tests) prior to release, according to the UX scorecard and the tracking process. Similarly, user support and escalation models were also optimized.
- **Validated user awareness and readiness:** Based on the UX scorecard, validation work of user awareness and readiness happened prior to WEA2 release. Communication materials were delivered (e.g., email) according to the TUX tracking process. Two surveys were conducted to check the progress of user awareness and readiness, and necessary actions were taken based on the feedback.
- **Conducted end-to-end TUX testing:** Unlike conventional usability testing, which mainly focuses on UI design, an integrated end-to-end TUX test was conducted with real end users across different job roles in a simulated environment that included call center desks (support scripts and agents), various help materials, and a back-end system support team. The end-to-end TUX test enabled the team to validate all the TUX touch points with real scenarios and various people who represented different roles in the business process. The test also gave the program one more chance to identify possible UX issues across all the TUX touch points prior to the release.

The WEA2 solution was released with great success. For instance, overall user satisfaction was increased from 43% (WEA1) to 78% (WEA2); the completion time for a major user task was shortened from >45 minutes to <20 minutes; the user-support call volume was decreased from 1.23 calls per 1,000 to 0.81 calls per 1,000. This case study demonstrates how the UXD solution was executed to address all TUX touch points through a streamlined business process, optimized UI design, enhanced user support model and training, and so on, resulting in an enhanced TUX.

4.1.3 Formalizing the UXD process

Figure 4 outlines the process of addressing TUX across a UX lifecycle. The overall process is highlighted below:

Step 1. Build a UXD team: The team consists of various TUX stakeholders who own individual TUX touch points, including people from UX professionals, training, communication, marketing, quality assurance, user support, and others. The team should report directly to the program management office.

Step 2. Conduct TUX gap and needs analysis: The TUX gap and needs analysis should reveal the TUX gaps in current use of products or services, not just in UI and usability but also from other TUX touch points. If there is no previous release, then the analysis should focus on user needs for current release. The analysis provides a foundation for defining UX requirements for subsequent UXD activities and the priority of efforts.

Step 3. Define a UXD scorecard and a tracking process: The UXD scorecard defines success criteria across all the TUX touch points, beyond conventional UI usability success criteria. A tracking process defines the time window to check the implementation of each TUX touch point and corresponding validation methods (e.g., surveys, usability tests).

Step 4. Execute and collaborate on UXD: The UXD team works together to execute the UXD process as planned, addresses issues around all TUX touch points, including business process, UI design, training, user help materials, user support model, communication, and marketing, among others. Take necessary actions based on the issues identified during each check, as defined in the tracking process.

Step 5. Conduct an integrated end-to-end TUX test: This test, unlike conventional usability tests that aim at UI design, checks all TUX touch points in a simulated environment where the real UX ecosystem is realistically presented as much as possible, including real end users, the product, user support desk, back-end tech support, and training materials, among others.

Step 6. Improve TUX and make end users ready for release: Based on the results of the end-to-end TUX test, the project team needs to make it a high priority to fix identified issues and ensure that end users are ready for the release.

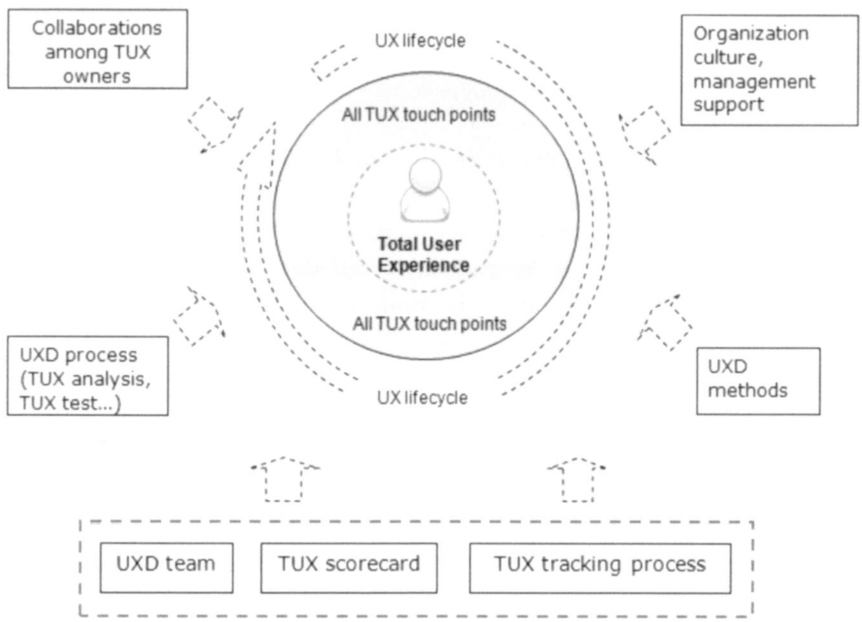

Fig. 4. The process framework of addressing TUX across a UX lifecycle.

Specifically, a UXD process is highlighted below in a specific UX ecosystem context of a corporation's IT setting, where typically off-the-shelf (OTS) products are purchased (Finstad et al., 2009; Sward, 2006). The overall UX ecosystem for an OTS product in this setting can be represented by five UX lifecycle stages as shown in Figure 5. Over the course of the UX lifecycle, end users interact with the product through all aspects of TUX. Relevant UX risks and key UX questions are outlined below for each of the UX lifecycle stages. Also, necessary UXD activities are suggested in order to address the corresponding UX risks and questions.

Stage 1. Marketing and user awareness: At this UX lifecycle stage, from a UXD perspective, the main goals are to clearly communicate the impending change and to set clear expectations to the end users so that they are aware of what is being delivered and what the impact will be. Questions for consideration: Is the end user aware of what is changing (and why)? Have expectations been set properly? Are communications targeted and timely? UX risk involves poor expectation management, confusion, and escalations. Possible UXD activities include surveying or interviewing end users to identify TUX issues in the previous release, if applicable, or their expectations and need for products with similar functionality, and conducting gap analysis by leveraging all available data (e.g., call center, email feedback) to identify UX gaps and needs. Thus, the project team can understand user needs for the upcoming release and better manage user expectations, and the users are ready to use the new product.

Stage 2. Order, delivery, and install: The main goal at this stage is to ensure that end users are able to successfully complete all tasks associated with initial usage of a solution without support. Questions for consideration include: Is the set-up process intuitive? Does the set up materials indicate whom to contact if help is needed? Risks include user frustration, inability to successfully complete a task, increased demand for support, escalations, and so on. Possible UXD activities include usability consulting, heuristic evaluation of the installation process, and usability testing of the materials. The project team also needs to ensure that the products to be purchased meet user needs and that they can be easily configured and installed. For OTS products, UX must be considered in the vendor selection process, including product TUX assessments and TUX scoring incorporated into the purchase decision-making matrix.

Stage 3. Product or service use: The main goal at this stage is to ensure that the product to be delivered is easy to use. Questions for consideration include: Is the product UI intuitive? Is the associated workflow easy to follow? Does the functionality meet business needs? Can the task be completed successfully with or without any help or support? Possible UXD activities include optimizing the business processes based on TUX data, optimizing product UI design through configuration changes if customization costs are significant, and collaborating with vendors to fix top UI usability issues, if identified.

Stage 4. User training and support: The main goal at this stage is to ensure that end users can easily and quickly receive support as needed. This is especially important to OTS products, where the UI design, the business process, and the configuration design may not be optimal for meeting corporate user needs. The product support can help mitigate potential UX risks left over from all other UXD activities. Questions for consideration include: Is the training effective for the user so minimal (if any) support will be required? Does support desk staff have documentation and training needed to support end-user needs? Are escalation and resolution paths clear? Possible UXD activities include collaborating with, training, and supporting business owners to jointly define a user-centric approach for delivering user- centric training and user support as needed.

Stage 5. Removal or End-of-Life: The main goals at this stage are to ensure that end users can successfully complete all tasks associated with EOL of a solution and/or seamless migration between an old solution and a new solution without support. Questions for consideration include: Does the shutdown or migration require additional support? Can the migration be completed with little to no manual intervention? Are instructions easy to follow? Do they indicate whom to contact if help is needed? Possible UXD activities include conducting an integrated end-to-end TUX test with real scenarios and actual end users in a simulated environment to test all TUX touch points, launching surveys to test user awareness and readiness, and launching effective communications based on the level of user awareness and readiness.

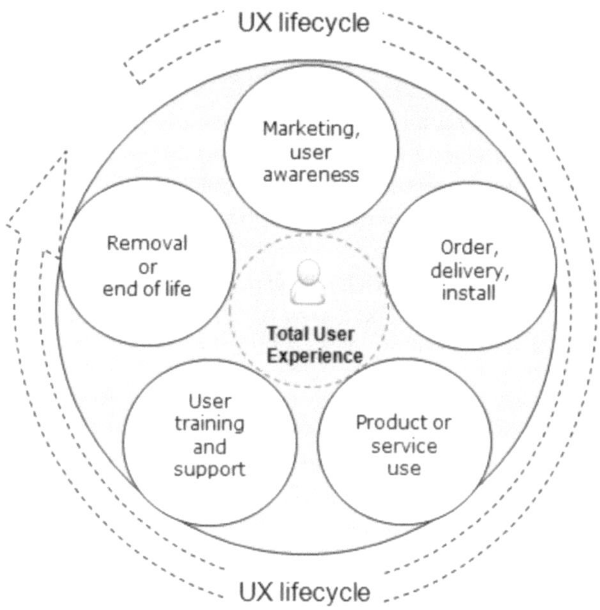

Fig. 5. The five UX lifecycle stages for an OTS product in a typical corporation IT setting.

4.2 Case study 2: Develop UX roadmaps to influence strategic directions of products

4.2.1 Problem statements

An internal business portal is a platform that provides corporate users with a collaborative, productive workspace by aggregating a variety of web content, applications, and reports. It allows users to access the content in a one-stop-shop approach based on their job roles with personalizable UX. Intel has been leveraging portals to enhance employee productivity. The UX problems in this case study come from two past projects. The first example has to do with a corporate business portal for internal financial users. The finance portal was released in the early 2000s with a personalization capability. The capability allowed users to turn some content on and off or move it around, similar to what iGoogle or myYahoo provide today. However, users felt frustrated when using the personalization functionality; they were not familiar with this type of capability, as there was no iGoogle or myYahoo at the time. The capability was eventually removed. This example shows that if a technology or business capability is ahead of UX and user readiness, it will not be accepted by end users and eventually will not deliver business values.

The second example has to do with an internal enterprise application. The product program was in the process of implementing a new version to replace the existing one. User research results indicated that users were not satisfied with the functionality provided by the new version (nor the existing one, for that matter). As compared with the UI design of a benchmark product with similar functionality on the market, the UI of the selected new version provided a lot of unnecessary data with less configuration flexibility. This would slow down the decision-making process. The program decided to add a customized UI presentation layer onto the new version of the application by using rich Internet application (RIA) technology, so that UX could be implemented that is similar to the benchmark product in today's market. This example shows that when technology or business capability lags behind user needs and UX, no one can deliver an optimal UX to end users.

The portal program had defined technology and business capability roadmaps for the next several years in order to enhance internal business portals to foster employee productivity. On the one hand, business and technology individuals are looking for predictable UX data to help guide their roadmaps to match the optimal UX sequence based on the lessons learned; on the other hand, the program had only the UX data that defined the current UX states (e.g., user needs, interaction models), which were typically delivered by a project HFE in terms of short-term user needs. There was no predictable UX data that could help the program optimize the proposed technology and business capability roadmaps.

4.2.2 The UXD solution

The business portal program requested that the human factors engineer (HFE) team to help identify UX gaps and needs, build the UX vision (near- and long-term), and define UI concepts for a next-generation business portal. While executing activities for these original goals, the team also leveraged the efforts to generate predictable UX data in terms of user needs and usages, so that the program could better plan its technology and business capability roadmaps accordingly, in order to deliver optimal UX over time, based on end user needs (Wooding & Xu, 2011).

The methods used in the study included: 1) industry best practice reviews (e.g., industry reports, external benchmarking), 2) information process mapping and observation sessions that allowed representative employees to map out the typical information and the workflow they use to support their daily jobs at Intel, 3) interviews with portal end users and observation sessions in their actual working environment to better understand their daily work patterns and usage models of the portal, and 4) a large-scale employee survey that collected their usage data on the portal and user needs for the portal from both near- and long-term perspectives.

The data analyses focused on identifying basic patterns and leveraging them to create larger patterns, and then looked for themes within themes; eventually the analyses led to the future UX vision, UX guiding principles, and user needs and the prioritized needs over time. Finally, a UX roadmap was created. Figure 6 illustrates the part of the UX roadmap concept that defines UX in terms of high-level usages in a subset area of the business portal domain.

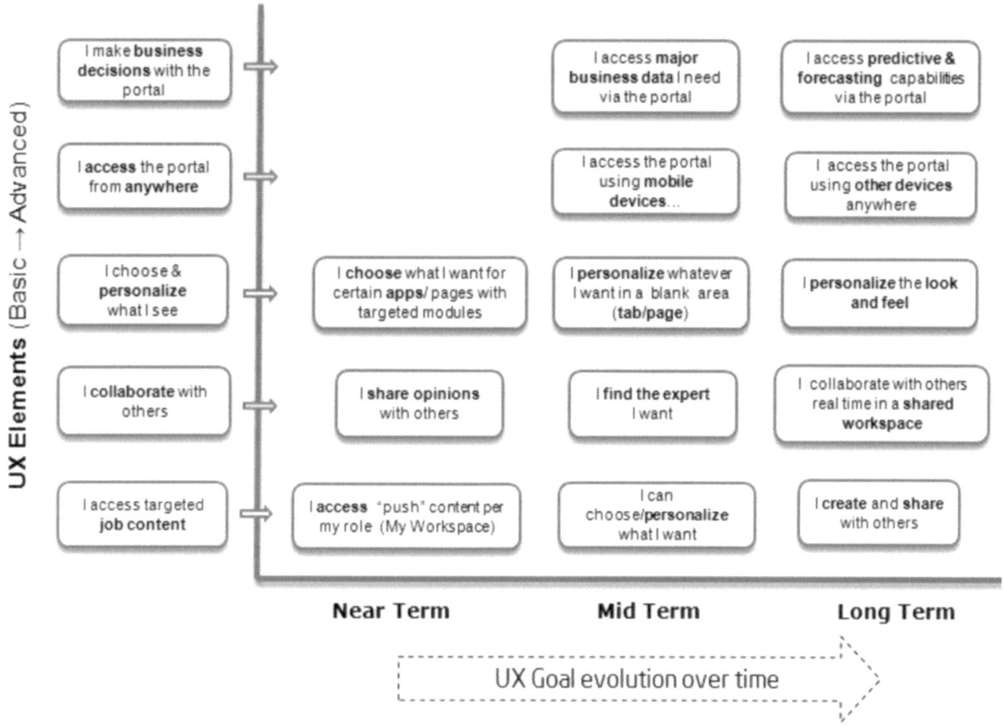

Fig. 6. Example of a UX roadmap for a business portal. Only high-level information is presented here for a subset area of the portal business domain.

Prior to creating the UX roadmap, HFE defined a UX vision and several UX guiding principles within the domain, based on the data collected. Overall, employees look for a business portal that not only provides information and news, but also provides more transactional data to help them take action and make business decisions. The collected data led to a UX vision; that is, employees can access information and do their work in one easy place with the portal. Following the UX vision, several UX guiding principles were defined. For instance, the portal should enable more collaboration, integration, and target content based on job roles; the portal content should be more relevant and personalizable. Put together, the UX vision and UX guiding principles helped shape the UX roadmap. Specific explanations for the UX roadmap follow:

- The vertical axis defines various UX elements in order, from basic to advanced. Basic user needs as defined by basic UX elements must be satisfied prior to advanced UX needs that will be satisfied later on. For instance, "I make business decisions with the portal" is the most advanced user need (UX element), but basic UX elements must be satisfied first, such as "I access targeted job content."

- The horizontal axis defines a sequence of sub-level UX goals over time, specifically for each of the UX elements. For instance, in order to achieve the "I access targeted job content" UX element, users need to access the "push" content per their job role first (i.e., the content is displayed to target users by default based on their job roles without user's touch); then to choose and personalize what they can access (a "pull" model). Eventually, the long-term goal, as defined by "I create or share content with others", will be achieved, which provides the capabilities for employees who want to do something beyond the push and pull models to facilitate collaborations with others.
- Figure 6 only presents a high-level view of the UX roadmap in terms of usages. A detailed view was also developed in terms of some near-term measurable UX goals. For instance, for the near-term UX goal of "I access 'push' content per my role," at the detailed level, UX goals were broken down into: 1) "I can access major job content by default with less than three clicks" for a project phase 1 deliverable, and 2) "I can access major job content by default with just one click" for a project phase 2 deliverable. Here, the measurable UX goals can be validated by project UX work during typical project-level UCD activities.
- Notice that no actual technology capabilities or product labels are defined in the UX roadmap above. A UX roadmap should only present user needs in a technical-agnostic way. Actual technology capabilities should be documented in a technology roadmap by mapping the UX roadmap and technology capability accordingly.

The proposed UX roadmap was presented to the product program with positive feedback. The program formed a team, including an HFE to further define the UX strategy for the program, and made adjustments of the existing technology capability roadmap by mapping both predictable UX data and technology capabilities accordingly (Chouhan et al., 2011). As a result, the sequence and the appropriate technology capabilities were optimized in a revised technology capability roadmap based on the optimal UX sequence, as defined in the UX roadmap. For instance, the implementation sequence of technology capabilities (e.g., corporate social media technology, enterprise workspaces technology) should be carefully defined in order to best satisfy user needs (i.e., "I collaborate with others") over time. In this case, some basic corporate social media capabilities must be implemented first (i.e., "Unified Employee Profiles"), and then additional capabilities (i.e., "Expert Finder") can be effectively utilized based on employee profiles. Eventually, the user need of "I find the expert I need" and then the user need for a shared workspace (i.e., "I collaborate with others in real time in a shared workspace") can be effectively satisfied in the long run.

This case study shows that a UX roadmap helped UX professionals document and communicate predictable UX data in a more influential way. Also, the UX roadmap helped technology and business people better understand UX and validate their roadmap; they were able to make necessary adjustments to ensure that both roadmaps were well aligned in order to deliver optimal UX over time.

4.2.3 Formalizing the UXD process

Figure 7 illustrates the process of developing a UX roadmap. The major steps are highlighted below:

Step 1. Gather data: Conduct user research to gather data on UX gaps and user needs in the relevant domain. The data may be documented in terms of usage scenarios, personas, etc. The data analyses should result in a better documentation of UX and user needs in terms of priority over time. The data gathered is the foundation for the development of a UX roadmap.

Step 2. Understand existing technology and business capability roadmaps: In many cases, existing business and technology roadmaps or strategies may be available. UX professionals should fully understand this information. The information provides great context for developing a UX roadmap. Also, an understanding of this information helps UX professionals effectively communicate with technology and business people.

Step 3. Define a UX vision and UX guiding principles: Analyze the collected UX data to generate a UX vision for the relevant domain. Collaborate with business and technology people to ensure that the UX vision is aligned with a long-term business strategy. Also, develop UX guiding principles based on the UX data; these principles will define the boundary and driving vectors for the UX roadmap.

Step 4. Build a UX roadmap: Draft the UX roadmap based on the data collected and roadmap conventions (see details in Section 3.2.2 of this chapter) with the guidance of the UX vision and UX guiding principles. Once drafted, review the proposed UX roadmap with business and technical partners (e.g., product manager, architect) to gather their input and revise it based on the feedback. This may be an iterative process.

Step 5. Manage influence: This is a key step in which UX professionals take opportunities to present the UX roadmap to various business and technical stakeholders, and work with them to develop or adjust (if already developed) their business and technology capability roadmap. There may be a lot of discussions and debates in this process. UXD professionals need to have UX data ready to support the discussions.

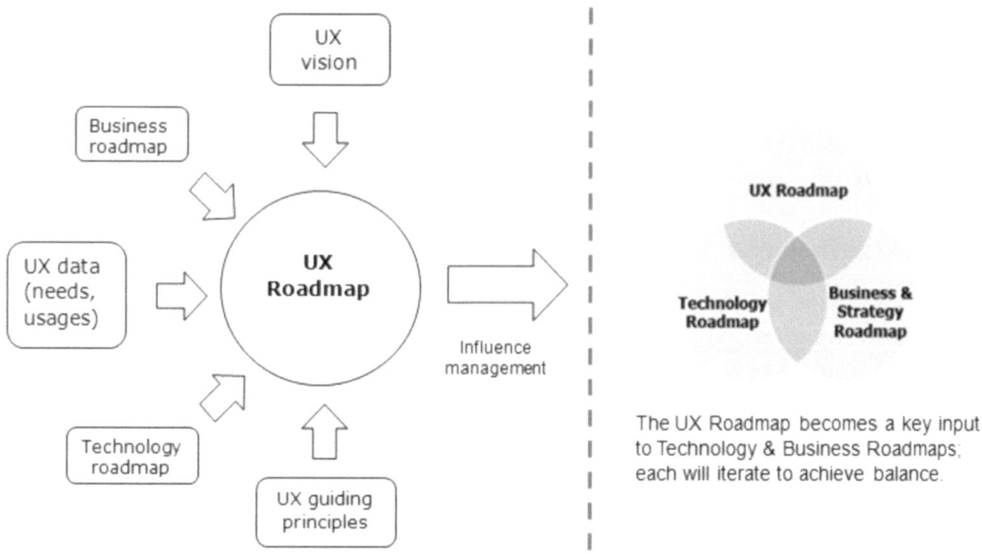

Fig. 7. The process of building a UX roadmap and influencing business and technology capability roadmaps, resulting in an optimal product roadmap that balances the needs across UX, technology, and business.

As a result, in order to achieve an optimal solution, one must consider the needs from all the three parts (technology capability, business capability, and UX). In reality, sometimes a trade-off decision may be needed. As Figure 7 shows (the right side of the figure), an optimal solution resides in the balanced, overlapping area across all three circles—technology, business, and UX. By doing this, an optimal product roadmap can be developed with balanced needs across UX, technology, and business.

4.3 Case study 3: Identify a new UX landing zone to influence the definitions of a new market opportunity and platform architecture capabilities

4.3.1 Problem statements

Popular TV technology failed to progress in the way other technologies had, leaving the living room with a comfortable void, since Internet experience, social networking, and contextual information are basically offered by other devices such as smartphones, netbooks, and laptops (Loi, 2011). Industry has been looking for new technological solutions and marketing opportunities for traditional TV technology. More specifically, the integration of Internet experience into traditional TV usage seems the most promising opportunity (Intel, 2011; Loi, 2011).

However, many UX-related questions remain open before a new UX landing zone can be validated (Intel, 2011): What kind of UX do consumers expect from the Internet access via a TV? How can one integrate the Internet experience while preserving the best of a TV medium which continues to inspire 1.3 billion households around the world? Will the worldwide reach of the Internet help link programming with an expanded pool of interested viewers? What is the best way to get content on viewers' radar screens? What type of interaction models do consumers expect? What are its implications to the TV screen design and user control UI design? What content do people want to watch and store? From a marketing perspective, what type of new market opportunity will this merge bring in if the new UX can be justified? In addition, what type of processing power and platform architecture capabilities are required to support the new TV experience and UI technology, based on desired interactions?

4.3.2 The UXD solution

To answer these questions, Intel UX professionals, along with other designers and technology people, has explored this new area in the last several years (Intel, 2011; Loi, 2009, 2011). As a result, a new TV experience and new market opportunity were defined; that is, the smart TV, which is called Google TV in the market. The smart TV allows users to access the Internet; to search online, personal content, and broadcast programming from a single TV interface; access downloadable apps; connect to social networks while watching favourite programs or movies; control TV with a unique new remote control or voice commands; and access an infinite amount of entertainment possibilities.

Related UXD activities are highlighted below to illustrate how these activities influenced the identification of the new market and the definition of the platform architecture capabilities (Intel, 2011; Loi, 2009, 2011).

- **Conducted early UXD activities:** In the past several years, Intel UX professionals, including anthologists and ethnographers, conducted a number of exploratory studies. They visited hundreds of people in their homes across India, the United Kingdom, the United States, China, and Indonesia to learn how they engaged with their TVs. These studies were aimed at various aspects of the TV experience and the social lives of television users. Unlike traditional user research (e.g., user groups, interviews), these studies intended to understand how people in various cultural settings touch the TV technology in their daily lives through direct observations and daily living with them. The UX professionals also conducted field studies on the retail floor (e.g., Best Buy) in America, where they talked with and listened to salespeople and consumers to understand what consumers needed from TV technology.
- **Identified user needs and modeled their usages:** The series of studies revealed overall user needs. For instance, they wanted to browse online while communicating and collaborating through social media while watching TV; they needed to access personal media on TV; they needed a way they could get whatever they wanted on demand. Furthermore, consumers wanted the UX quality of this new technology to be simple and interactive. Overall, a new usage model is clearly emerging: the intersection of television and the Internet.
- **Conceptualized the UX:** Based on the collected data and identified usage models, the UX professionals partnered with interaction designers, architects, and other technical people to define new UX concepts for various UX components through UI prototyping. These UX components include home media aggregation, TV widget (rich Internet apps), a 3-D UI, the ability to share/send personal content with/to others or to access/receive contextual information and recommendations, gesture-based navigation, and voice-based search.
- **Validated UI concepts:** Numerous usability tests were conducted to iteratively assess and improve these proposed interaction models and UI concepts through quantitative and qualitative UX assessment metrics. During the iterative process and interactive discussions among UX professionals, interaction designers, marketing people, and technical people, these concepts also deeply influenced people's thinking and the development approach.
- **Influenced platform architecture definitions:** The newly identified UX and usage models, along with the support data from both qualitative and quantitative UX data, helped open up new opportunities with internal stakeholders (e.g., architects, product owners) to conceptualize their technological frameworks. It eventually influenced the definitions of platform architecture. As a result, an Intel CPU was designed specifically for powering the smart TV (Intel, 2011). The CPU offers platform capabilities to help design a usable smart TV, such as home-theatre quality, audio/video performance, signal processing, surround sound, 3-D graphics, and etc.

In summary, the deliverables through these efforts met corporate strategic marketing needs and also provided a reference design for Intel when Google approached the company looking for hardware and platform solutions for Google TV. It opened a door for smart TV, which is not just a product but rather a completely new product category of TV (Lois, 2011). In addition, the UX professionals' early involvement in the first stages provided a UX foundation for the platform architecture capabilities, which enabled OEM (e.g., Google, Sony) to develop usable products to meet consumer needs.

4.3.3 Formalizing the UXD process

Figure 8 illustrates the process framework of the UXD solution. Major steps are highlighted as follows:

Step 1. **Gather UX data:** Conduct field user research with target users through methods such as ethnographic study and contextual inquiry. In contrast to conventional user studies, these types of studies should be conducted in a broad social-tech environment, and UX professionals may work or live with users to gain a deep understanding of their needs and usages of emerging technology.

Step 2. **Define usage models:** Analyze collected data to build usage models. Usage models define users' real needs, values, and the interaction environment by describing product usages and context. The usage models also tie several product development artefacts together around UX, including architecture, key features, requirements, and technologies. Eventually, the usage models help drive detailed UX definitions.

Step 3. **Identify new UX landing zones:** This is one of the key steps where UX professionals collaborate with other partners (e.g., architecture, marketing people, and business owners). A UX landing zone may be defined in a minimum, target, and outstanding format, which helps define a market opportunity for a new product by satisfying end user needs in terms of priority. Eventually, a UX landing zone, in alignment with market requirements, is created for influencing the definitions of product requirements and platform architecture capabilities. In this way, UX is well integrated and fully considered in the early strategic planning stage, even before a product development process officially kicks off.

Step 4. **Conceptualize future UX:** At this stage, detailed UX can be developed based on the previous work through methods such as use cases, workflow, and contextual diagrams, eventually leading to interaction models and UI design concepts. The efforts in this process provide visualized materials (e.g., UI concepts) to communicate and document future UX.

Step 5. **Validate concepts:** To validate the UX, usability testing should be conducted for the proposed UI concepts with novel scenarios and usages. The UI may include both software and hardware UI. This may require iterative tests and improvements of the proposed interaction models and UI.

Step 6. **Manage influence:** Influence management also is a key step and should actually be an ongoing effort throughout the whole UXD process. UX professionals need to collaborate with various stakeholders, including marketing, technology, and business people, to influence their capability roadmaps, platform architecture definitions, and marketing opportunity definitions.

5. Future research

As discussed above, although a common ground is shared between the UXD approach and current UCD practices at a high level, the UXD approach is beyond the UCD approach in terms of processes and methods. The UXD approach is still under development. Thus, more research is needed in order to make UXD more mature.

First of all, UXD involves great collaborations across a variety of TUX owners across all the TUX touch points in the context of a UX ecosystem. Conventional methods in some areas

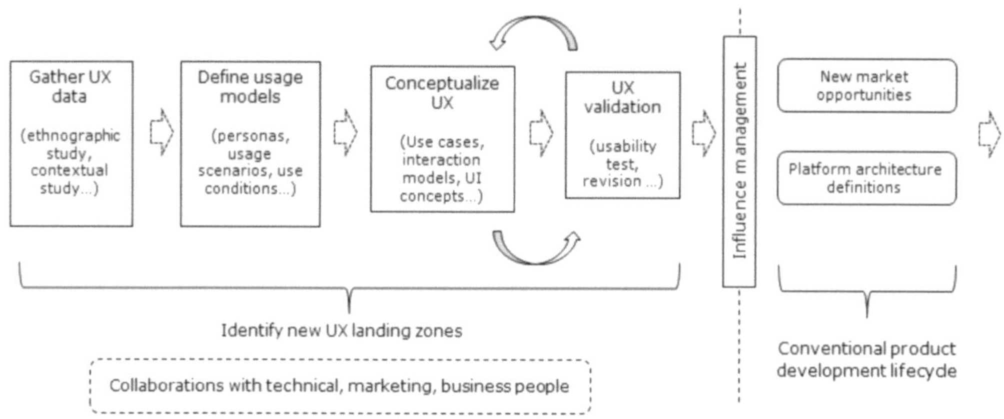

Fig. 8. The UXD process to identifying a new UX landing zone that influences the identification of a new product market and the definition of platform architecture capabilities.

(e.g., marketing, training, and business processes) need to be enhanced or integrated with UXD methods in order to support UXD activities more effectively, including more user-centric marketing and training methods, UXD success criteria definitions in these individual areas, and validation of success.

Secondly, more formalized and effective UXD methods need to be developed in order to support development of UX roadmaps, identifying emerging usage models and new UX landing zones in order for UX professionals to more effectively influence new market opportunity definitions and platform architecture definitions. For instance, methods for modelling usages of technology and modelling of UX in both quantitative and qualitative ways.

Also, a UXD process should be reasonably flexible to fit a variety of UX ecosystems in terms of scale and nature. New UX ecosystems are continuously emerging, and new components are being added to existing UX ecosystems, such as social computing and cloud computing. All of which make UX richer and more versatile. More best-known methods should be developed to help UX professionals address UX in a variety of UX ecosystems in today's dynamic and versatile social-tech environment.

Finally, UX is no longer an isolated experience within an individual platform, such as desktop computers, tablets, and smartphones. Computing technology is entering a "compute continuum" era, where computer resources (e.g., content, data, processes, and applications) are shared across different platforms. For instance, people want to access the same application across a smartphone, a tablet, and a desktop computer with seamless UX. This creates new needs and challenges to the continuum of TUX. The conventional consistent UI design principle is no longer feasible across platforms due to different platform UI conventions. The form factor between smartphones and desktop computers will definitely drive inconsistent UI. Here, achieving "UX continuum" with a consistent UX becomes a more important design goal, so that users can receive seamless UX across platforms without interruptions in different usage situations. This definitely expands the boundary of a UX ecosystem and drives new needs for UXD practices.

6. References

Bowles, C. & Bowles, C. (2010). *Undercover User Experience Design (Voices That Matter)*. ISBN-10: 0321719905, ISBN-10: 0321719905, New Riders Press.

Chouhan, H., Manohar, M., & Xu, W. (2011). User Experience Design (UXD) Framework and Principles: Content, Collaboration, Portal, and Search. *Intel IT Technical Report.*

Finstad, K; Xu, W.; Kapoor, S.; Canakapalli, S. & Gladding, J. (2009). Bridging the gaps between enterprise software and end users. *Interactions*, Vol. XVI.2, March + April, pp.10-14.

Intel Corporation (2011). *Consumer Electronics – Smart TV* http://intelconsumerelectronics.com/Smart-TV/

Loi, D. (2009) Leading through design enabling: practical use of design at Intel. *DesignConvexity– 8th International Conference of the European Academy of Design*, Aberdeen Scotland, April 1-3.

Loi, D. (2011). Changing the TV Industry through User Experience Design. *Proceedings of 14th Human Computer Interaction International*, pp.465-474, ISBN 978-3-642-21601-5, Orlando, FL, USA, July 9-14, 2011.

Nielsen, J. (1993). *Usability Engineering*. ISBN 0-12-518406-9, Academic Press, Boston.

Norman, D. (1999). *The invisible Computer: Why Good Products Can Fail, the Personal Computer is So Complex, and Information Appliances Are the Solution*. MIT Press.

Sward, D. (2006). *Measuring the Business Value of Information Technology*, ISBN 13: 978-0 - 976483-27-4, Intel Press, San Clara, CA.

Unger, R. & Chandler, C. (2009). *A Project Guide to UX Design: For user experience designers in the field or in the making*, ISBN-10: 0321607376, New Riders Press.

Vredenburg, K; Isensee, S; & Righi, C. (2001). *User-Centered Design: An Integrated Approach.* ISBN 0130912956, Prentice Hall.

Wooding, L. & Xu, W. (2011). User experience and architecture from end-user research to UX roadmaps. *Intel Enterprise Architecture Summit (2011).*

Xu, W., Dainoff, M., & Mark, L. (1999). Facilitating complex tasks by externalizing functional properties of a work domain on the user interface. *International Journal of Human-Computer Interaction*. Vol. 11, No. 3, pp. 201-229.

Xu, W. (2001). Integrating user-centered design approach into software development process: improving usability of interactive software. *Proceedings of 9th International Conference on Human-Computer Interaction* (the abridged proceedings), New Orleans, Louisiana, USA, Aug, 5-10, 2001, 127-129.

Xu, W. (2003). User-centered design: opportunities and challenges for human factors practices in China. *Chinese Journal of Ergonomics,*, Vol.9, No.4, pp 8-11.

Xu, W. (2005). Recent trend of research and applications on human-computer interaction. *Chinese Journal of Ergonomics*. Vol. 11, No. 4, pp 37-40.

Xu, W. (2007). Identifying problems and generating recommendations for enhancing complex systems: Applying the abstraction hierarchy framework as an analytical tool. *Human Factors*, Vol. 49, No. 6, pp. 975-994.

9

Higher Efficiency in Operations Can Be Achieved with More Focus on the Operator

Per Lundmark
ABB AB
Sweden

1. Introduction

In the early days of industrial automation, system designers attempted to automate everything and remove the human operator - whom they considered the weakest link in the process control loop - entirely. Today, it is clear that the human operator is an integral part of any automated control loop in almost all industrial applications of any size. Understanding and maximizing collaboration between the control system and the human operator is therefore essential. Furthermore, a systematic design approach to this task is crucial for reasons of safety and optimum system performance (Pretlove & Skourup, 2007).

The operational phase of any project is typically the dominant part of the total life cycle. Therefore, it is logical to focus on the operational efficiency and economical aspects. The global process industry loses $20 billion, or five percent of annual production, to unscheduled downtime and poor quality. ARC Advisory Group (www.arcweb.com) estimates that almost 80 percent of these losses are preventable and that 40 percent are primarily the result of operator error (Woll et al., 2002).

We can easily understand that our increased demand for higher productivity, better quality and increased safety has changed the situation for the operator over the last fifty years. More complex applications, more data to interpret and more alarms to process are some factors that affect the operator. With this increased responsibility for overall profitability and lack of continuous training, it has become harder to find operators willing to accept this burden and devote their working life to the control room. ARC Advisory Group estimates that most companies spend less than 2% of available hours on training (Wilkins, 2007). To make things even worse, operators are not always in focus when new control rooms are built. Lack of understanding of human factors, too much emphasis on technology and not enough involvement by operators in the planning phase of the control room all result in poor ergonomics and dissatisfied staff (Nimmo, 2007 and Ericson et al., 2008).

If we examine the consequences of this attitude, we find a very high employee turnover rate among operators. What's more, the costs of hiring and training new personnel are considerable. It is estimated that the cost of training one plant control room operator is at least $100,000 (Wilkins, 2007).

A growing problem is also the fact that many of today's operators are approaching retirement and it is difficult to recruit replacements among today's younger generation. This is particularly acute within the Mining and Oil & Gas industries, where many of the production sites are located in very remote and unattractive areas. ARC Advisory Group estimates that almost half of the operators retiring before 2030 can not be replaced (Wilkins, 2007).

It is time to bring the control rooms to a new level where operator effectiveness is highest on the priority list. With today's technology it is possible to consolidate control rooms into control centers offering completely integrated solutions. It is possible to work with sound, colors, lighting, intelligent furniture's, smart textiles and micro ventilation to achieve much higher efficiency in the operations than ever before. Furthermore, the Distributed Control System (DCS) of today fully support integration of power- and process automation in one common environment together with support for safety applications and advanced alarm management.

Our challenge is to create an attractive, safe and effective environment with operators in focus. Questions like "why do we need a control room, what tasks are to be executed by whom and how can we implement an operator interface that works safely even in critical situations," must be asked. "How can we build the most impressive display wall for our visitors and how much money can we save by buying non-ergonomic furniture and skimping on good control room layout planning" are aspects that should never be raised.

2. Existing issues in many control rooms

Many sub-optimal issues can be identified in existing control rooms. The most obvious are listed here.

- Operators do not have a good overview of the complete process
- The control room environment is not optimized for the actual number of operators
- The Human Machine Interface (HMI) is not optimized for operator tasks
- Large displays are not implemented with the operator in focus
- Close Circuit TeleVision (CCTV) and Telecom equipment are poorly integrated in the operator environment
- The control room was built with limited focus on human factors and ergonomics
- The control room was not built for consolidation and collaboration
- No one is designated to be responsible for the total control room solution

2.1 There has always been a need for an overview

Let us take a look at some history. There has always been a need for an overview. Before the operator control room was available, the operator had to walk around the process and smell, feel and listen to the different parts of the plant. The first attempts to support the operator implied that all instruments, switches, etc. were gathered at one common location. See Fig. 1. Information and interaction were combined in the same piece of hardware. A switch could, for instance, be moved in different positions with direct feedback on the current status.

Fig. 1. Early attempt to create a plant overview.

The next step was the development of chart recorders, alarm enunciators and single-loop controllers mounted in large wall panels. It was now possible to get a very good overview of the process with recorded trends, differentiated alarms and loop status. Of course, all interaction could be carried out directly at the wall panels. See Fig. 2.

Fig. 2. Wall panels with a good overview and full interaction.

As computers were developed, it became possible to move the wall panels onto several process graphics with full interaction. However, the new problem created was that the total overview was now lost. Each operator screen became merely a keyhole into the process. See Fig. 3. Navigation was another subject for improvement. With only one screen (or possibly two), it was difficult to find the required information and act in a timely fashion. In many installations, this was solved by adding absolutely everything possible to one single screen, thus avoiding the need for display navigation. The problem with this solution is obvious. If something critical happens on this screen, it is difficult to interpret the information in a secure way.

Fig. 3. Each operator screen is acting as a keyhole into the process.

Fig. 4. Large display walls make it possible to present a plant overview for the operators.

The solution to this shortcoming with computer screens and HMI software was large display walls. Finally, here was a way to replace the wall panel with an electronic version that could display the total process overview with support for modifications. It was even possible to use part of the large display wall as a CCTV monitor. Large display walls like these are still commonly found, particularly in control rooms in sectors like Oil & Gas and Utilities. See Fig. 4. Unfortunately, these large display walls have shown to create problems when designing a good layout in the control room. This is further discussed in section 2.4 below.

The latest solution to all of the above shortcomings is the interactive personal large display integrated with the operator console. In a 2011 report, ARC Advisory Group provides the Extended Operator Workplace (EOW) from ABB as a good example of this new type of console (Woll & Miller, 2011). See Fig. 5. The EOW is designed for highly complex 24/7 operation. A large display is mounted behind the normal monitors. All parts of the console, including the large screen, are motorized for optimum working conditions. The large screen is completely interactive for safe, fast and correct decision-making. This means that faceplates, trend information, operator instructions, maintenance records and any other object-related information can be accessed on the large screen, in the actual context, without any delay or need for separate browsing.

Fig. 5. Extended Operator Workplace from ABB with interactive large display.

2.2 Control rooms must be designed for the operator needs

Many times the control room is designed based on the actual process, not necessary with the needs of each operator in mind. The specification for the DCS system, Request for Proposal (RFQ) is often structured based on the different process sections in the plant. Each section is typically given an operator workplace with one or two monitors. When all this is combined on the operator console, we can typically see 8-16 monitors with many separate keyboards and mouse arrangements. In addition to these process monitors, the operator is exposed to several CCTV monitors, telecom equipment and other supporting systems. In total, each operator can easily be overwhelmed with information and different devices to interact with the process and other people. See Fig. 6.

When designing the control room, it is important to start from the correct number of operators and the different tasks at hand. Who needs to be in the control room? How many monitors can each operator handle? How many keyboards are needed? Is it possible to add one or two additional operators during critical situations? How much of the information can be integrated to avoid separate monitors and keyboards? These are all questions that must be asked. If there is an expansion to the operation, it is very important to go back to these questions instead of just adding one or several more operator workplaces with yet one or two more monitors for each process section that is added. The reality is that no more operators are added just because the operation is expanded.

Fig. 6. The operator can easily be overwhelmed with too much information.

2.3 The HMI can be improved with high performance design

It is obviously important to use appropriate fonts, colors, shapes etc. when designing the HMI for operators. It is equally important to have all relevant information integrated to be able to work effectively in a consistent way. It is more important for the operator to easily read the actual information than to know where and how to find it. Operator workplaces should be possible to personalize for optimal operation for different individual needs. In today's advanced operator workplaces with many connected display channels and support for large displays, it is important to have a good navigation system and a predictive behavior for different categories of information. The operator should be able to focus on the information, not on moving and resizing windows. It is also important that open windows for different categories of information are reused to avoid overload with too many open windows on the screen. It should of course be possible for the operator to control the behavior and override any preconfigured rules.

2.4 Large display walls can make it difficult obtain a good control room layout

If the decision to use a large display wall is not coordinated with the operator needs, the resulting control room layout might reduce the possibility to achieve an optimal control room operation.

Unfortunately, large screens are not always implemented with the individual operator in focus. The main purpose of a large screen is to present an overview of the total process for everyone in the control room, with the emphasis on deviation from the normal process state. As soon as such a deviation is identified, the operator has to move his/her focus to the normal screens and translate the relative deviation to something measurable in real numbers. This can be very stressful, especially with many other people in the control room hanging over their shoulder.

Yet another problem is the way the display wall affects control room layout. The wall ends up defining the layout of the complete control room, and it thus limits possibilities for future changes. Valuable floor space is wasted on both sides of the wall. Space behind the wall is needed for maintenance access, but there is also a recommended minimum distance between the wall and the operator consoles. This latter space is normally used as a walkway that generates disturbing traffic in the control room. The fact that the display wall is fixed in position also makes it difficult to have adjustable consoles. For example, a large display wall mounted to allow consoles to be adjusted for standing operation would be too high for operators who prefer to sit. See Fig. 7.

Fig. 7. Large display walls have a great impact on the control room layout.

2.5 It is essential to integrate CCTV and telecom equipment

In critical situations, it is not effective to change focus from the task at hand, and try to find the actual CCTV monitor showing a specific process object or area.

There are typically two ways to implement CCTV in the control room. The most commonly used approach is to add a CCTV monitor for each camera. These monitors are then typically hanging down from the ceiling or positioned on the wall. See Fig. 8. A more sophisticated solution is to use a dedicated operator station with a camera switch and a joystick. The operator can from this station select one or several camera images and control each individual camera with the joystick. The obvious problem with both these approaches is that the operator must change his/her focus from the actual process and keep unnecessary information about what camera to select and how to move it into position. An other, more serious problem is that the operator has no means to look at recorded information in any easy way.

Fig. 8. CCTV and Telecom equipment should be integrated with the DCS system.

The only acceptable solution to CCTV is to integrate the functionality into the DCS system. This way, the operator can get to the camera information, when needed, without knowing where the camera is located. It should of course be possible to operate the camera (Pan, Tilt and Zoom) from the DCS system and easily retrieve recorded information in a similar way as looking at trend displays. If needed, any video window should be possible to share with any other operator connected to the same network. This should even include field operators with wireless handheld terminals. ARC Advisory Group wrote a report in June 2010 that emphasized the importance of integrated real-time video. *"Integrated real-time live video into human machine interface (HMI) tools provides an excellent opportunity to maximize operator effectiveness and ergonomics…"* (Resnick, 2010).

If the live video is combined with audio communication, it is also possible to turn the DCS system into a video conferencing system. This would instantly turn the operator console into a true collaboration center where process specialists and field operators together can solve complex situations.

The same applies to Telecom equipment that must be integrated with the DCS environment to avoid unnecessary movements in the control room. Critical alarms should for instance be possible to broadcast over the Telecom system in different languages directly from the DCS system without loosing the focus from the process. Selected alarms and events should also be possible to automatically distribute over email and Short Message System (SMS). See section 2.7 for more details about the consequences of control room consolidation.

2.6 There must be more focus on ergonomics and human factors

The control room environment must be designed for the operators, not for the technical equipment. It is a known fact that we are all different. We have different length, we have different preferences when it comes to sitting or standing, we have different vision with different requirements when it comes to lighting, distance to the screens etc. We have also very different perception of temperature. Some operators like it a little bit warmer, while other operators like it a little bit cooler. Without focus on the operator environment, it is obvious that we easily are introducing problems in the control room. If these problems are causing personal injuries and avoidable sick leave or unexpected turn around, we really need to look for a better solution. See Fig. 9.

Fig. 9. It is important to focus on ergonomics and Human Factors.

Today, it is possible to find operator desks that are motorized and height adjustable from 650 mm up to 1300 mm. The freedom to vary posture is known to be beneficial for operator health and thus effectiveness. The monitor boards used for regular wide screen monitors can be individually motorized and height adjustable from +70 to -130 mm independently of the working board. All wide screen monitors, on the monitor board, can be tilted simultaneously from +5 degrees down to -45 degrees through a unique motorized monitor support. The distance between the monitor board and the working board can also be motorized and adjustable up to 150 mm. Achieving a perfect viewing distance and angle is at all times easily secured with this sophisticated arrangement. See Fig. 10.

It is also possible to integrate a large overview screen with the console and that way avoiding wasted space on either side of it. Disturbing traffic is eliminated and the consoles can be moved around as conditions change. Furthermore, as the consoles are ergonomically designed they can be adjusted for individual operators. It is not far fetched to imagine that the console adjusts to individual preferences as part of the login process and then automatically adjust to various situations during the working shift.

Fig. 10. The operator console must be ergonomically designed to fit different people.

To minimize the need to move around in the control room between different computer screens and CCTV monitors, there are keyboards that can serve several computers. This way, it is possible to operate an advanced operator console without leaving the operator chair and lose the focus on the process. It is also possible to have directed speakers located above the operator to secure that the operator can hear alarm sounds or background music without disturbing the other operators in the control room.

All computers should normally be removed from the control room and placed in a separate computer room with a controlled environment. In this way, the noise level can be kept to a minimum, and it is much easier to keep the control room floor clean. All of these factors work in favor of attracting new and hopefully younger operators into the control room.

The latest trend is to control the ventilation system in the control room to create different temperature zones around each operator. This is an efficient way to save energy as the temperature can be individual adjusted. It is even possible to connect the temperature control to the DCS system such that certain alarm levels can trigger more cold air and keep the operator alert in critical situations.

2.7 It is important to understand the consequences of consolidation and collaboration

There is a clear trend today to consolidate many separate control rooms into intelligent control centers. This way we can utilize the expertise from several experienced operators in one common center that can be remotely located from the actual plant(s). See Fig. 11.

With today's technology, it is possible to operate remotely without any visual contact with the actual process. It is possible to utilize integrated CCTV and telecom equipment to communicate over long distance. It is also much easier to recruit operators in an attractive area compared to living in a camp far away from the nearest city.

When consolidating control rooms, more operators from different parts of the process, must be able to work side by side collaborating in an efficient way. There are technical solutions available today that make it possible to shower the operator with personal sound without disturbing the other operators in the same room. With this technology, it is even possible for the operators to listening to their personal favorite background music.

As discussed earlier, it is essential that that the operators can get all relevant information from one single location in the control room. This implies that all information must be integrated in the DCS system to avoid unnecessary movements in the control center. If several operators need to see the same information e.g. a CCTV image, the information can be sent by one operator to other operators for sharing. This even includes field operators with wireless handheld terminals.

One other important aspect of working in a collaboration center is that you must be able to solve critical situations in a group. Therefore, it is crucial that the operator console can be operated to standing position to allow all involved operators to work on the same level. To have several people standing behind you looking over your shoulder can be very stressful in a critical situation where you can not afford to make a wrong decision.

Fig. 11. Several experienced operators in a consolidated control center.

2.8 Someone must be responsible for the total control room environment

There are many challenges during the total life cycle of a control room project. The process is expanded with more process sections, the existing computer monitors are getting obsolete, changed standards are requiring modified safety measures etc. Without someone being responsible for the total control room, the environment will very quickly decay when many different vendors are implementing different systems without coordination. The result from this are different types of monitors, different screen resolutions, different viewing angles many different keyboards and other input devices. See Fig. 12. Even if the actual operator is used to this situation, it can be devastating if someone new without the experience is exposed to this environment in a critical situation.

Fig. 12. Someone must be responsible for the total control room environment.

Today, there are many good tools available that simplify the planning of a control room with support for change management. It is important that we take control room planning serious over the whole life cycle of the operation. See Fig. 13.

3. CPAS and operational excellence

ARC Advisory Group introduced a new vision for Collaborative Process Automation Systems (CPAS) in 2002. The intention was to create an environment in which everyone could access all relevant data in context in a secure way. According to ARC, the definition of the HMI part of Operational Excellence is *"A single unified environment for the presentation of information to the operators as well as the ability to present information in context to the right people at the right time from any point within the system"* (Woll et al., 2002). See Fig. 14.

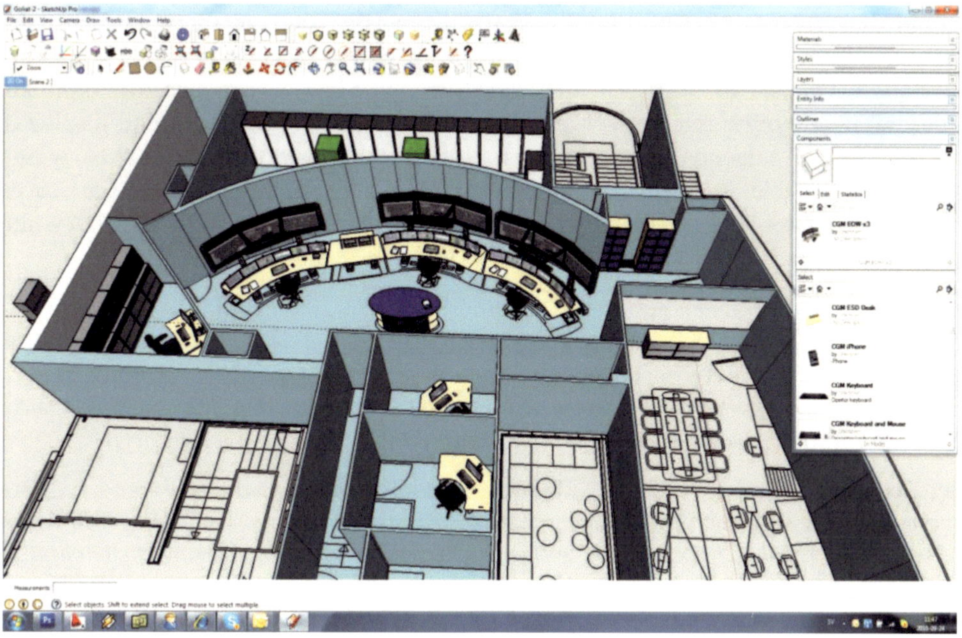

Fig. 13. Free tools like Google SketchUp make it easier to plan the control room.

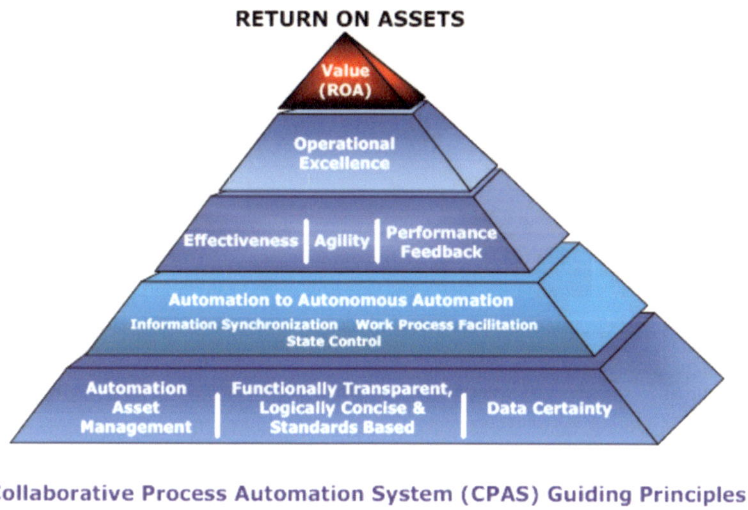

Fig. 14. ARC Advisory Group CPAS Guiding Principles.

What ARC means is that the operator environment has to be in focus if maximum Return on Assets (ROA) is to be achieved. The operator must have access to all relevant data and tools that help him/her make decisions and act quickly in relation to a situation in the process. All data must be synchronized and presented in a unified way, in context, and without the need to login and browse in separate systems. Navigation must be quick and intuitive to avoid delays when searching for data. Once again, we need to think about how we use a mixture of large screens and normal monitors. The large screen has to be interactive to allow for immediate display of critical information with tools to act. It is also important that all screens support transfer of information. If, for instance, an operator finds something important that he/she must share with others, there must be a way to send this information (duplicate) to any other screen (workplace) in the system. It could be a trend display that must be shown on someone's large display for further investigation, or a live video window that must be possible to see on multiple screens, even over a long distance. (It can be very limiting for operators if a video window is presented in the corner of a display wall without being able to move it or duplicate it to any other location on any other screen.)

ARC Advisory Group also emphasizes the importance of ergonomics in the control room. In a report written in July 2007, it recommends that *"Design and implementation of control room and HMI, should include ergonomics and change management"*. It is further recommended that *"Technology providers should ... propose solutions and implementation approaches that include ergonomics and change management skills"* (de Leeuw, 2007).

In a report written in October 2008, ARC Advisory Group introduced a new trend. *"... This is part of a trend that ARC refers to as "Ergonometrics", where increased ergonomics leads to increases in KPIs and metric results. The objective is to offer the operator an attractive working environment with extended functionality, which better enables functional consolidation and increases collaboration. The key component of this offering is the Extended Operator Workplace, which provides detailed overview images of the entire process with high definition graphics. This solution creates an optimized working environment for the operators and meets high ergonomic standards, making it more possible for the operators to act fast in critical situations and avoid expensive shutdowns"* (Resnick, 2008).

Fig. 15. Operator environment with focus on ergonomics and human factors.

Operational excellence and operator effectiveness means a lot more than just functionality in the DCS system. Ergonomics and focus on human factors are equally important to keep the operator alert, healthy and ready to act. So what do we mean with ergonomics and human factors? Let us repeat some of the most important factors that affect the operator in the control room:

- Physical factors like: working height, viewing angle, legroom and sitting comfort
- Ambient factors like: lighting, noise level, temperature, humidity and air quality
- Number of screens per keyboard and resolution on different screens
- Lighting and color depending on process state (smart textiles and daylight control)
- Sound systems for public and personal information
- Traffic control (field operators, visitors and other control room personal)
- Access to other functions or rooms (printer room, rest room, kitchen, toilet, meeting room, offices, computer room, library, exercise room, emergency room etc.)
- Console proximity (communication and collaboration)

See Fig. 15 for a modern operator working environment with focus on ergonomics and human factors.

4. Conclusion

The fact that we need to change the way we plan control rooms and move from a technology-driven approach to one that is operator-focused is quite obvious. We need to create a safer and securer environment that will attract new operators into the control room. To do this, we need to challenge ourselves. Are there other new ways to plan the control room? Are there other new technologies that allow us to think differently than before? Can it be that the younger generation has different demands and requirements?

Let us look at an example. In a typical bid specification, all operator seats are specified as two-monitor seats with no information about any human factors or how these seats are supposed to fit into the control room layout. Sometimes we see a specification for a large display wall with a number of projection cubes and an overall size. What are missing in these specifications are the reasons behind these numbers. Why only two monitors per operator seat? Why should the display wall have a certain size? What is the purpose of the large display wall? What information will be presented on the monitors and what on the wall? How should information be presented and how should the different screens and monitors interact with each other?

This bid specification typically illustrates a technology-driven approach. We start with the known hardware facts without any thoughts about the operators and soft factors.

For the next control room project, therefore, we should change our focus from technology and cost-fixation to a complete focus on the operator and the total control room solution. With this approach, we will find that it is possible to combine higher productivity with better quality and safer operation. Operators will be more satisfied and we will lower the turnover of new workers.

5. Acknowledgment

I would like to thank the following companies and organization for their contribution to this chapter about Operator Effectiveness.

ABB Corporate Research for their contribution with ideas of how to create the operator environment of the future.

CGM/FOC for their contribution with pictures and ideas of how to create the operator console of the future.

Chalmers University for their contribution with research projects to find the optimum operator environment for more alert operators in critical situations.

ARC advisory Group for their contribution with several reports in the area of Ergonometrics; the teamwork between human operators and technology.

6. Nomenclature

CCTV	Closed Circuit TeleVision
CPAS	Collaborative Process Automation Systems
DCS	Distributed Control System
EOW	Extended Operator Workplace from ABB
HF	Human Factors
HMI	Human Machine Interface
RFQ	Request For Quotation
ROA	Return On Asset
SMS	Short Message System

7. References

de Leeuw V. (2007). Improving HMI Design to Optimize Operations. ARC Advisory Group

Ericson M., Strandén L., Emardson R. & Pendrill L. (2008) FDC prestudy- State of the art survey of control rooms. SP Technical Research Institute of Sweden

Nimmo I. (2007). Human Factors design of control room environments. User Centered design services Inc, USA. Lectures arranged by HFN, Swedish Network for Human Factors

Pretlove J. & Skourup C. (2007). Human in the Loop, The human operator is a central figure in the design and operation of industrial automation systems. ABB Review, The corporate technical journal of the ABB Group. 1/2007, page 6-10

Resnick C. (2008) Briefing: ABB's Extended Operator Workplace Focused on Ergonometrics. ARC Advisory Group

Resnick C. (2010) Real-time Video provides a Fourth Dimension for HMI Software. ARC Advisory Group

Wilkins M. (2007). Human Factors Best Practices for Automation - Addressing the Skills Gap. ARC Advisory Group

Woll D., Hill D. & Polsonetti C. (2002). Collaborative Process Automation Drives Return on Assets. ARC Advisory Group

Woll D. (2006). Briefing: ABB's Extended Operator Workplace Projects a Perfect Picture. ARC Advisory Group

Woll D., Miller P. (2011). Improving Operational Performance by Improving the Operator Experience. ARC Advisory Group

10

Critical Thinking Skills for Intelligence Analysis

Douglas H. Harris and V. Alan Spiker
Anacapa Sciences, Inc.
USA

1. Introduction

Whether performed by national agencies or local law enforcement, the ultimate objective of intelligence analysis is to develop timely inferences that can be acted upon with confidence. To this end, effective intelligence analysis consists of integrating collected information and then developing and testing hypotheses based on that information through successive iterations of additional data collection, evaluation, collation, integration and inductive reasoning. The desired end products are inferences that specify the who, what, when, where, why and how of the activity of interest and lead to appropriate actions. This process is illustrated in Figure 1.

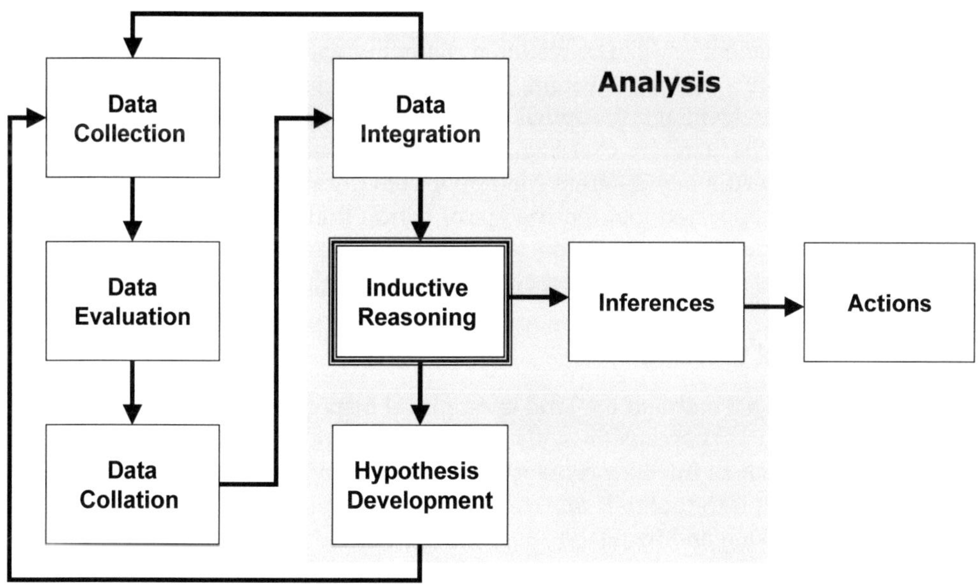

Fig. 1. The intelligence process

While in the last couple of decades a number of useful tools have been developed to aid in data collection, evaluation, collation and integration, analysis remains highly dependent on the cognitive capabilities, specifically the critical thinking skills, of the human analyst. For this reason, it is important to understand the inherent capabilities and limitations of the analyst and, in particular, the cognitive challenges of intelligence analysis that must be overcome through training in and application of critical thinking (Harris, 2006a, 2006b; Heuer, 1999; Moore, 2007).

Our concern with and study of critical thinking skills for intelligence analysis relates to that aspect of ergonomics research that seeks to understand how people engage in cognitive work and how to develop systems and training that best support that work. These efforts have come to be known as cognitive ergonomics or cognitive engineering. While our focus here is specifically on the domain of intelligence analysis, we recognize the many areas of endeavor that require critical thinking skills. These include the professions, business, military, education, and research and development.

Just what is critical thinking? Critical thinking was first conceived in the early 1940's by two psychologists, Goodwin Watson and Edward Glaser. Watson and Glaser also developed the first test of the skill, the Watson-Glaser Critical Thinking Appraisal (Watson & Glaser, 1980), which is still widely used. Since then, almost all of the theoretical development has been conducted by educators and philosophers, where the focus has been on identifying people with superior critical thinking aptitudes through testing. The notion of critical thinking as a *skill* that can be improved through focused training, as is the view of a psychological construct, has received far less attention. However, see Halpern (1996) and Baron and Sternberg (1986) for notable exceptions.

In desiring to develop a consensus definition, the American Philosophical Association attempted to develop such a definition based on the responses of 46 experts (American Philosophical Association, 1990). The resulting definition was "purposeful, self-regulatory judgment which results in interpretation, analysis, evaluation, and inference, as well as explanation of the evidential, conceptual considerations upon which that judgment is based." A review of the literature covering the 10 years subsequent to that exercise (Fischer & Spiker, 2000) revealed many different conceptions of critical thinking with only a modest degree of overlap. It appeared that the concept of critical thinking could not be adequately addressed by a simple verbal definition. A more comprehensive model was required to address important components and interactions, and to serve as a basis for empirical testing.

2. Model of critical thinking

Critical thinking has not endured the kind of empirical inspection typically bestowed upon constructs developed by psychologists. Its relationship to other, well-established psychological constructs such as intelligence, working memory, and reasoning, for example, has rarely been studied. It is the authors' admittedly subjective opinion that the lack of empirical study of critical thinking and its relationship to other individual difference dimensions has produced a fractionated view of the construct. Without the grounding of data, theorists have been free to postulate divergent concepts. An effort in philosophy to reach a consensus definition in 1990 had little effect on unifying the field.

To fill this gap, Fischer and Spiker (2004) developed a model that is sufficiently specific to permit empirical testing. The model identifies the role of critical thinking within the related fields of reasoning and judgment, which have been empirically studied since the 1950s and are better understood theoretically. It incorporates many ideas offered by leading thinkers (e.g., Paul & Elder, 2001) in philosophy and education. It also embodies many of the variables discussed in the relevant literature (e.g., predisposing attitudes, experience, knowledge, and skills) and specifies the relationships among them.

The model can, and has been, used to make testable predictions about the factors that influence critical thinking and about the associated psychological consequences. It also offers practical guidance to the development of systems and training. An overview of the model's main features is provided here following a brief review of current thinking about reasoning and judgment, on which the model is based.

2.1 Dual system theory of reasoning and judgment

Prior to the early 1970's, the dominant theory of decision making stated that people made judgments by calculating (1) the probability and (2) the utility of competing options. Although this rational-choice model took on a variety of forms, all versions posited a rational actor who made calculations of probability and/or utility, and selected the option that had the highest value. In the 1950's, however, researchers began to notice that the model failed to predict actual behavior (Meehl, 1954; Simon, 1957). Evidence that falsified the rational choice theory accumulated over the following decade.

In the early 1970s, an alternative theory proposed that people use heuristics, as opposed to the rational weighing of relevant factors, to make judgments. The "new" theory was, and continues to be, supported by empirical study (Baron & Sternberg, 1986). The heuristic theory states that many judgments are based on intuition or rules of thumb. It does not propose that all judgments are made intuitively, just that there is a tendency to use such processes to make many judgments. The most recent versions of heuristic theory, in fact, propose that two cognitive systems are used to make judgments (Kahneman, 2003). The first system, intuition, is a quick, automatic, implicit process that been proposed to explain judgment. To accommodate the multiple theories, many researchers now use associational strengths to arrive at solutions. The other system, reasoning, is effortful, conscious, and deliberately controlled. Since the 1970's, multiple and similar two-process theories have referred to the implicit associational type of process as System 1, and the conscious deliberate process, as System 2. The following example shows how these two processes may lead to different judgments.

Suppose a bat and a ball cost $1.10 in total. The bat costs $1 more than the ball. How much does the ball cost?

Most people's immediate judgment is that the ball costs 10 cents. This is a response derived from intuition or System 1, which again, is quick, automatic, and relies on associations. The strong mathematical association between $1.10, $1, and 10 cents leads to this quick, but wrong, judgment. The ball can't cost 10 cents because then the bat would have to be $1, which would make it only 90 cents more than the ball. The more effortful deliberately controlled reasoning, or System 2, process usually produces a different, and correct, answer. When people spend the time and effort to think about the problem, they usually realize the ball must cost 5 cents and the bat must cost $1.05. Hence, in this example, the two systems produce different judgments. It would be a mistake to conclude that System 1 always produces different judgments than System 2, however. Nor does System 1 always produce an incorrect answer, nor one that is poorer than one produced by System 2.

In fact, researchers have shown that expert performance in any field, which is commonly the gold standard, is often driven by intuition derived from extensive experience (e.g., Klein, 1999). That said, expert performance is not without fault, and studies have shown that even experts make errors in judgment when well-learned associations lead them astray (Thaler & Sunstein, 2008). The associational processes used in System 1 that make expert performance so quick and powerful are the same processes that are responsible for systematic errors that experts sometimes make. Additional weaknesses of System 1 are that it depends on the quality and amount of experience an individual possesses, and it can't be used effectively in novel situations. System 2 reasoning also has its strengths and weaknesses. While it is highly useful in novel situations and problems, it is also slow and effortful. It usually cannot be utilized concurrently with other tasks and, like System 1, it can also produce wrong judgments.

Most recent theories, however, believe that Systems 1 and 2 *run in parallel* and *work together*, capitalizing on each other's strengths and compensating for their weaknesses. For example, many researchers believe that one function of the controlled deliberate process is to monitor the products of the automatic process. System 2 is thought to endorse, make adjustments to, correct, or block the judgment of System 1. However, if no intuitive response is accessible, System 2 may be the primary processing system used to arrive at a judgment. The similarities between descriptions of critical thinking and System 2 are striking. The words "effortful, controlled, deliberate, purposeful, and conscious" are frequently used to describe both.

2.2 Overview of the model

As shown in Figure 2, the model assumes that critical thinking skills are executed by System 2, and that these skills also serve to monitor, evaluate, and control the judgments produced by the System 1 associational process. Hence, Figure 2 shows that System 1 judgments provide input to critical thinking skills. The two processes are thought to run in parallel and interact to produce judgments. Because System 1 is truly an automatic and uncontrolled process, it cannot be consciously initiated or stopped. For this reason, only the products, and not the process, of System 1 is monitored. Because System 1 is quick, it often comes to judgment before System 2, but System 2 may override, or confirm, that judgment. Therefore, System 2 has the potential for controlling judgment, although it may not always utilize that potential.

Critical thinking can provide a thorough examination of the problem at hand. Although System 1 might derive just one solution (Klein, 1999), System 2 can provide multiple potential solutions. System 1 works to narrow possible action paths, which is often highly effective when the task must be accomplished quickly and when the problem space is limited. However, when the problem space is novel or complex or when solutions must be innovative, critical thinking skills are more powerful. They also have the meta-cognitive capability to monitor the progress of their own processing, as represented by the self-monitoring arrows leading out and back into the System 2 processor in Figure 2.

Figure 2 also shows how the processing engines interact with environmental and individual factors. Both systems receive initial input from the environment in the form of information about a situation or problem that requires judgment. Part of that input is a meta-task that defines the general purpose of judgment. The other part of the input is information about the situation. System 1 immediately and automatically begins processing of the input by searching through its associational network for potential solutions that will satisfy the purpose. Critical thinking, motored by the System 2 processing engine, receives the same input, filtered through predisposing individual difference factors, which are discussed in greater detail below. If critical thinking skills are engaged, they will begin to evaluate solutions offered by System 1 or they will apply deliberate reasoning to the problem.

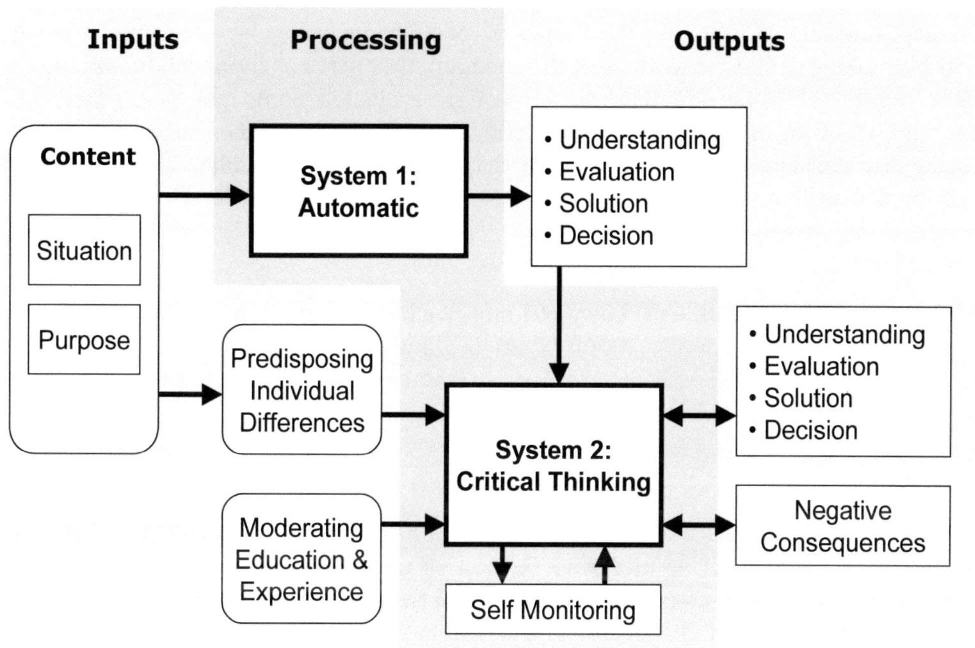

Fig. 2. Model of critical thinking (from Fischer & Spiker, 2004)

Whether or not critical thinking is utilized depends on a variety of factors, including individual predisposition and situational variables. The sum value of these factors provides the impetus to engage in effortful critical thinking, but that motivation must exceed some threshold value. In the paragraphs below, each component of the model is examined in more detail.

2.3 Components of the model

2.3.1 Inputs

As noted above, the opportunities for judgment are set in motion by the contextual factors--the situation and the purpose. While the automatic System 1 will engage in all conditions, two characteristics of the situation must be present to elicit critical thinking: the stimulus material must contain substantive information and there must be sufficient time available to engage System 2. Other characteristics of the situation that make it more likely that System 2 will be engaged include the presence of conflicting information, disordered or unorganized material, uncertain information, and complex material .

Critical thinking is not an end in itself, but serves objectives specified by purpose (meta-tasks). The purpose also dictates the specific response that will be required to successfully end the process. For example, the situation may include a meta-task to understand, make an evaluation, make a decision, or solve a complex problem. Even if the final result is based on System 1 processing, System 2 determines when the requirements of the purpose have been met. Hence, successful completion of the meta-tasks as determined by System 2 can also provide input that terminates an episode.

Predisposing factors influence the likelihood of a person using, or persisting in using, a critical thinking skill. Like features of the situation, they serve as input conditions, and as a filter through which the situation and purpose are evaluated. Some may be key factors that strongly affect an individual's use of a critical thinking skill. Other factors may have a weaker relationship to critical thinking, perhaps increasing the likelihood of engaging in a skill by a marginal amount. In summary, predispositions are measurable ways in which people differ, whether fixed or modifiable, that influence the use or persistence of use of critical thinking.

Moderating variables influence how, and how well, critical thinking skills are performed. For example, domain expertise, recent experience, and education influence the quality of the reasoning produced by the process. They do not, however, influence whether one executes a particular skill, as do predisposing factors.

2.3.2 Processing

The task posed by a particular situation should not be confused with the system that is used to solve it. For example, one may have the task of understanding an intent statement that could be achieved using associational processes of System 1 or controlled skills powered by System 2. Therefore, an individual who is trying to understand an intent statement may or may not be using critical thinking to do so. Even more important, the application of critical thinking skills driven by System 2 does not always produce the best solution to a task. It would be a mistake to encourage the exclusive use of critical thinking because that strategy would deny the power and effectiveness of System 1. Similarly, it is not advisable to only develop associational processes because controlled deliberate reasoning can both produce superior solutions and provide necessary checks on the products of System 1. Moreover, the issue of which system is most effective is practically irrelevant because most theorists believe that both are almost always used in conjunction to produce a solution. Hence, the real issue that determines the quality of a solution is how well the two systems interact.

There is a general consensus in the literature that individuals are reluctant to engage in critical thinking (Moore, 2007). This is based on widespread observation of incoherent reasoning, nonsensical beliefs, lack of respect for evidence, poor reasoning test scores, and unsupported decision-making in various populations. Indeed, much of the literature is devoted to a movement to increase the application of critical thinking in various populations. One of the central topics has been the question of why the public seems disinclined to use it. Some theorists posit that individual characteristics, such as intellectual laziness, arrogance and cowardice (which are represented in the model as predisposing individual differences), are the reasons why it is avoided. The model of critical thinking discussed here, however, posits that negative affective consequences associated with the application of critical thinking are the primary inhibitory sources.

The model posits that individuals who engage in critical thinking for any substantive length of time are likely to experience negative affective reactions. For example, the process can produce mental fatigue, increased effort, increased anxiety, cognitive dissonance, and decreased self-esteem. Negative affect experienced during an episode might be countered by positive affect that is the result of a positive outcome (e.g., solving a difficult problem) that, in turn, is a direct result of critical thinking. Therefore, its application can be positively rewarded and hence, increased use may be realized. Some individuals, then, may not experience associated negative affect; but at the very least, by definition, critical thinking requires more effort than System 1 processing, and is therefore a less desirable means to achieve judgment in that limited sense.

2.3.3 Outputs

The quality of a solution produced by the application of a critical thinking skill is likely to be affected by how well the skill is executed. Decrements in performance may be produced by failing to apply an essential component (e.g., failing to clarify ambiguous information in a message or failing to consider alternative explanations for a pattern of data), failing to perform accurately a component of the skill, or by lacking sufficient knowledge to be processed. Therefore, one could apply critical thinking and still produce inferior solutions to a task. Moreover, it is not possible to determine whether System 1 or System 2 was applied to derive a solution based on the solution alone. The quality of a solution may also be affected by moderating variables such as educational level and experience. These issues are important to the design of training that seeks to improve critical thinking skills.

Figure 2 shows that negative experiential consequences serve as both a byproduct of critical thinking and as input to the decision to maintain a critical thinking episode, as depicted by the bidirectional arrow. When the affective consequences of applying the critical thinking skill become too negative, the motivation to maintain the episode is decreased. If the negative consequences are sufficiently strong, they may result in a cessation of the episode.

Finally, it should be recognized that effective critical thinking depends on gaining insights as well as reducing mistakes (Klein, 2011). Critical thinking is valuable for reducing mistakes but, in the process, may interfere with the process of gaining insights. It is notable that the concept formulated by the American Philosophical Society (1990) encompassed both reducing mistakes (by analyzing arguments, assessing claims, querying evidence and justifying procedures) and enhancing insights (by decoding significance, examining ideas, and conjecturing alternatives).

2.4 Validation of the model

Some preliminary research has been completed toward validating the model (Fischer et al., 2009). A series of controlled studies was conducted of the effect of web-based critical thinking training on the information interpretation and analysis performance of Army officers. Subjective responses from the participants indicated that the training was considered highly relevant, beneficial to their military work, offered training that was not available to them elsewhere, and that the self-paced feature of the program was highly desirable.

Objective measures indicated that the training encouraged critical thinking and enhanced the understanding and analysis of information that resulted from a greater depth of processing. This was evidenced by increased officer sensitivity to likely errors, increased awareness of weak elements that might easily be overlooked, and by an enhanced ability to distinguish between information actually present and their own inferences about or interpretations that go beyond the information explicitly provided. Participants who completed the critical thinking training made significantly fewer unjustified inferences than participants assigned to the control conditions; they did make inferences but justified them by pointing out explicit supporting information. Therefore, the training appeared to encourage discrimination of what is "known" or "given" from what might be inferred.

3. Human limitations that affect critical thinking

Our experience to date in training and applying intelligence analysis skills suggests that some of the principal challenges that affect critical thinking are human limitations. Humans are limited in their capabilities to address complexity, by the biases they bring to the process, by their difficulties in handling uncertainty and, often, by the lack of relevant domain expertise (Harris, 2006a, 2006b; Heuer, 1999).

3.1 Complexity

The complexity of information to be analyzed can increase rapidly and easily. For example, from calculations of combinations, there are 6 possible ways that 4 entities can relate to each other but there are 496 possible ways that 32 entities can relate to each other. The potential extent of complexity becomes apparent when one realizes that it is not uncommon for an analyst to address hundreds or thousands of entities. Since it has been well established that humans' ability to process information is greatly constrained due to working memory limitations (Miller, 1956; Baddeley, 1986, 1996; Engle & Kane, 2004), complexity can be a significant analytical challenge. Of course, there are various other contributors to complexity—types of relationships, variability of conditions, and so on (Auprasert & Limpiyakorn, 2008). Moreover, some of the simplifying strategies that analysts might employ may lead to biased results, such as focusing on vivid, immediate cases rather than on more abstract, pallid statistical data that are often of much greater value.

3.2 Bias

There are also many ways that bias can affect the analysis of information (Heuer, 1999) but, for the intelligence analyst, combating confirmation bias is one of the greatest challenges. Confirmation bias is the selective use of information to support what we already believe, ignoring information that would disconfirm the belief. Examples of tendencies most humans share that contribute to confirmation bias are:

- humans tend to perceive what they expect to perceive and, as a consequence, valuable experience and expertise can sometimes work against an analyst when facing new or unexpected information or situations;
- mind-sets are quick to form but resistant to change, leading analysts to persist with a hypothesis in the face of growing disconfirming evidence; and
- well-established thinking patterns are difficult to change, leading to difficulties in viewing problems from different perspectives or understanding other points of view.

3.3 Uncertainty

The work of the intelligence analyst is conducted within the realm of uncertainty and with the aim of reducing the veil of uncertainty through which judgments, decisions and actions must be taken. Since few inferences in the dynamic, complex world of decision-making lend themselves to the rigor of statistical analysis, most of the objective, mathematical approaches to the assessment of uncertainty are not applicable. Thus, in assessing and communicating the level of confidence that should be associated with a specific inference, the analyst must employ subjective conditional probabilities. That is, not only must critical thinking skills be employed to assemble evidence, generate premises and develop an inference, they must also be employed to arrive at the level of confidence one should have in the inference (Klein et al., 2006).

Moreover, the analyst is faced with a tradeoff between the level of detail in an inference (the answers to who, what, when, where, why and how questions) and the level of confidence that can be given to the inference. More detail provides a more useful inference but typically at the sacrifice of confidence; less detail provides a greater level of confidence but typically at the sacrifice of usefulness. One of the challenges faced by the analyst is to make an effective tradeoff between detail and confidence.

3.4 Domain expertise

The final potential problem, to be discussed here, for the intelligence analyst is the lack of domain expertise; that is, an analyst cannot be expected to be an expert in all of the information domains required for a typical analysis. Critical thinking skills are required to compensate for lack of domain expertise and, also, to facilitate the development of expertise in domains that are important to current and future analyses. Closely related to this challenge is the availability of information, which might range from large volumes in some domains to very little in others. In the first case, critical thinking is required to sort out the relevant from the non-relevant from the volumes available and, in the second, to develop assumptions to be used in place of non-available facts. Another problem is language, where analysts may have to depend on translations away from original sources or where cultural information is vital to the analysis but they don't have much prior knowledge of the culture.

4. Challenges ahead for intelligence analysis

At the 2006 annual meeting of the International Association of Law Enforcement Intelligence Analysts, the US Deputy Director of National Intelligence for Analysis described his view of the challenges ahead. His main point was that the extension of current trends (for example, increased globalization, communications flow, opportunities for terrorism) will continue to blur the line between personal security and national security, which in turn, will blur the line between law enforcement and military operations, and between activities involving people and those involving territory (Fingar, 2006).

There is increasing awareness of the importance of intelligence, particularly that from open sources. A senior advisor to the Secretary of Defense recently stated that most information (perhaps as much as 90%) that matters now is available to anyone with an internet connection, that understanding and influencing foreign populations was very important, and that future enemies are unlikely to confront the world's overwhelming military power with conventional warfare, but with a technology-assisted insurgency (Packer, 2006).

Open source intelligence is an intelligence-gathering discipline that involves the collection, analysis, and interpretation of information from publicly available sources to produce "usable" intelligence. It can be distinguished from research since the former's intent is to create tailored or customized knowledge to support a particular decision or satisfy a specified information need by an individual or group. The sources of this information are now quite vast, and include media (newspapers, magazines, radio, TV, Internet), social networks (Facebook, Twitter, YouTube), public data (government reports, speeches), observation and reporting (plane spotters, satellite imagery), professional and academic (conferences, papers), and geospatial dimensions. The latter are often glossed over, but must be considered since not all open source data is text-based. These data come from various sources, including maps, spatial databases, commercial imagery, and the like. As information has become more available by virtue of the Internet and other digital media, the physical collection of information from open sources has become much easier.

5. Application of available technology

Technology is now employed extensively by intelligence analysts to extract meaning from available information, to support the performance of a variety of analyses, and to aid in the communication of analytical results to the users of intelligence. The design of systems to support the intelligence process, and specifically intelligence analysis, can benefit from what we now know about the nature and role of critical thinking in this process. This knowledge of the specific skills required also supports the application of cognitive ergonomics to the development of training systems and methods that best meet analyst performance requirements. To be meaningful and realistic, training content and exercises must be developed and implemented within the context of available technology. Below, we summarize some of the technology that might be employed for extracting and analyzing information, the design of which can benefit from cognitive ergonomics that addresses specific critical thinking skills.

5.1 Extraction of entities, concepts, relationships and events

Software applications are required to analyze, from any source of text data, and automatically extract many different entity types, such as people, dates, location, modes of transportation, facilities, measurements, currency figures, weapons, email addresses, and organizations. The extraction capability is extended to the detection and extraction of activities, events and relationships among these types of entities. Automatically extracting this information means that analysts do not have to read extensive amounts of text to pull out these types of information manually; they can focus sooner on the relevant information. Automated event and relationship extraction helps analysts more quickly discover associations, transactions, and action sequences that can be employed in the development of link, event and activity analyses. Therefore, assuming that this is done effectively, analysis can begin with information that has been automatically extracted and organized from much more voluminous amounts of information available to the analyst.

Information relevant to global operations might be in various languages other than English, such as Arabic, Chinese, Farsi and many others. Technology is available to support and augment the efforts of the limited number of translators typically available to exploit foreign language documents. Language processing software can help translators analyze documents in their native language and help them select the most relevant documents or sections of documents for translation. Available software might contain a suite of natural-language processing components that enable language and character encoding identification, paragraph and sentence analysis, stemming and decompounding, part-of-speech tagging, and noun phrase extraction. With such a system, analyst training can assume that the capabilities exist to provide the analyst with information that has been extracted and translated relatively effectively, by means of automated and human processing, from numerous different languages.

Software can also provide user-guided text extraction from unstructured data sources, supporting the transformation of user-identified text-based information into structured graphic formats for further analysis. The user can highlight important information contained in text documents—entities and associations among entities, for example—and easily put it in chart form to enhance visualization of the information without having to retype information. This type of conversion can be employed with a variety of text formats and applications.

5.2 Database development and query capabilities

Technology can also help store, organize and query data extracted from multiple sources. Multi-user databases can now be built relatively quickly without the need for advanced, specialized technical expertise through the use of built-in forms and automated importation of information from data extraction tools and systems. Complex database query languages that previously had to be learned by analysts can now be replaced by simple, more intuitive, ways to query data, such as using graphics to "draw" questions. Some of the tasks that can be facilitated by currently available technology include the following:

- Conduct full text searches of the database to find exact matches, synonyms or words that sound similar to those in one's search criteria.
- Draw the query question by dragging and dropping relevant graphic icons and links from previously constructed charts. One can then save, organize and share queries and information with other analysts.
- Reveal all relationships between a selected chart item and other entities in a database.
- Visually establish the shortest path between two data elements, even if the relationship involves several degrees of separation.

- Maintain the quality of the database by searching a set, a query result or the entire database for duplicate information.
- Create reports that can be printed, posted to a web page, or saved in a word-processing application to facilitate the communication of query results.
- Enable location-based database queries by interfacing with available geographical mapping software.
- Interface with analytical software to provide the means for allowing the manual analysis of data and/or the automatic generation of charts, such as link diagrams, event timelines and financial transaction flow charts.

Geographic information system technology and services are available to augment database development and query capabilities. For example, required geographical information can be obtained through a web-based map interface (e.g., Google Maps, Google Earth, Ushahidi), providing access to geo-referenced infrastructure data. One existing system provides more than 1,300 layers of infrastructure data encompassing the physical, economic, socio-demographic, religious, health, educational, energy, military, transportation, political, governmental, geographical and chemical infrastructures of the United States. For example, some systems can provide the name, address, administrator contact information, number of beds and personnel for each hospital in the United States. Similar information can be provided for schools, fire stations, airports, and related facilities.

For the part-task training exercises and scenarios required to develop critical thinking skills, database development and query capabilities are not likely to be required of the trainee. However, the development of training exercises and scenarios, to be realistic, must be compatible with current and future database configurations, formats and capabilities. For this reason, the training developer must be knowledgeable about these and future systems and how they are likely to be employed in the intelligence process.

5.3 Data integration support

Analytical software applications now serve to support the analysis function by providing tools that permit the analyst to convert information into a variety of formats, from multiple sources, into graphic products that lead to greater understanding of the information by both the analyst and the ultimate user of analytical products. This is the part of the intelligence process that is typically referred to as data integration. Significant advances have been made in the development and improvement of these systems; further enhancements can be made through the application of cognitive ergonomics, specifically through the application of our knowledge about the critical thinking skills that must be supported.

Analysts can uncover and interpret relationships and patterns hidden in data through the generation of intuitive charts. Moreover, information about each entity and link portrayed on a chart can be accessed through embedded data cards connected to the displayed icons or through links from icons back to the database. A sample chart is shown below in Figure 3. The mechanics for obtaining the additional information is typically just a matter of clicking on the icon of interest.

One valuable capability that can be provided by analytical software applications is data filtering. An important critical thinking strategy to counter the effects of complexity is that of determining specific analytic objectives and filtering out information in the database that is not relevant to meeting that objective. Examples of specific analytical objectives include the following: defining the flow of money into a specific organization; clarifying the span of control of a specific individual; including only information above a specified level of validity; tracking events that occurred only during a specified time period; and examining financial transactions above a specified amount during a specified time period.

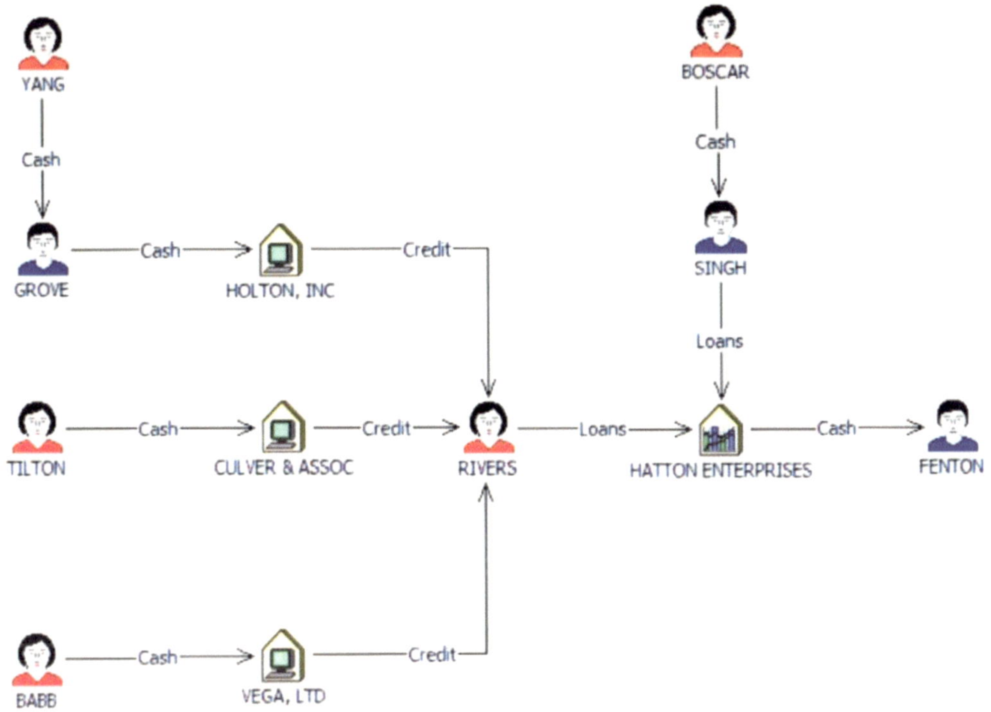

Fig. 3. Sample data integration diagram

The results from pursuing these specific objectives might provide support to a set of premises that lead to the development of an inference about the who, what, when, where, why and how of the activity of interest. Other capabilities provided by analytical software include the following:

- Switch between network and timeline views to identify patterns in both time and space.
- Automatically compare labels, types, attributes, names and aliases when combining data from different sources.
- Augment charts by including visuals such as maps and photographs.

6. Key critical thinking skills for intelligence analysis

Harris (2011) reviewed the literature and identified 120 elements considered by researchers and educators as important for critical thinking. Like elements were grouped together. Two survey instruments were then developed based on the listing of 18 critical thinking skills and designed to identify those skills that would provide the highest training payoff. The first instrument was designed to collect data from a sample of 73 intelligence analysts at a software user's national conference in Washington DC following a 60-minute presentation on critical thinking. The second instrument employed a similar, expanded approach to collect data from six instructors who conduct intelligence analysis training and 14 students who had just completed a two-week course on intelligence analysis. Analyses of these data identified 11 critical thinking skills that appeared to have the highest payoff for intelligence analysis and mapped these skills to four specific intelligence analysis functions:

- assess and integrate information,
- organize information into premises,
- develop hypotheses, and
- test hypotheses.

He then developed specifications for the development of web-based training on these skills, and developed and installed on-line prototype demonstrations of a critical thinking strategies overview module and a module for one of the 11 specific skills—consider value-cost-risk tradeoffs in seeking additional information. The 11 critical thinking skills are listed and mapped to intelligence analysis functions in Figure 4. A description of each skill is provided below, related to the intelligence analysis function it serves.

6.1 Assess and integrate information

The three skills associated with this first function are: envision the goal (end state) of the analysis, assess and filter for relevance and validity, and extract the essential message. These skills are described in the paragraphs that follow.

6.1.1 Envision the goal (end state) of the analysis

This skill is the ability to envision the desired goal (the desired end state of the analysis in terms of providing a useful inference that can be acted on with confidence in a timely manner) and to use that vision to guide and limit the analysis to tasks that will achieve the desired goal. This critical thinking skill constitutes an overall check on the process and products of thinking to ensure that it is moving the analysis forward along the right path.

There are many circumstances and reasons why an analyst might head down the wrong path, particularly early in an analysis. The directions given at the outset for conducting the analysis might be vague and confusing; the volume of information might be so great as to provide many opportunities to head in the wrong direction; and some types of information might be more compelling than others, even if not as helpful in meeting the analytical objectives. Consequently, particularly early in the data collection and integration efforts, the analyst must expend effort to envision the goal of the analysis and maintain that vision during the analytical process.

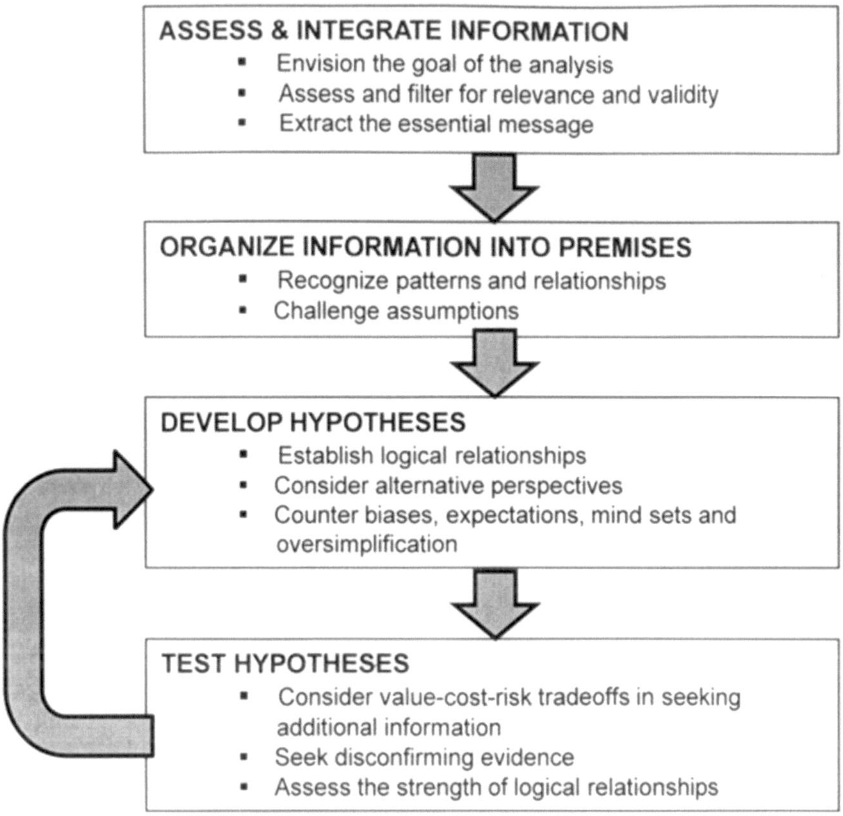

Fig. 4. Critical thinking skills grouped into the intelligence analysis functions they support

6.1.2 Assess and filter for relevance and validity

Critical thinking is required to distinguish between relevant and irrelevant information, and valid and invalid information, relative to the desired end state, purpose or goal of the analysis. This skill is obviously related to envisioning the goal, because the analyst needs a well-defined goal before being able to determine what information is likely to be relevant in meeting that goal. The principal skill involved here is the assessment of information for its potential relevancy to the objectives of the analysis; once relevancy has been determined one must then assess validity to provide assurance that it will contribute positively to the analysis.

Assessing and filtering information contributes to intelligence analysis during the assessment and integration stage. If one of the objectives of the analysis is to determine the relationships among entities of various types (for example: individuals, organizations, places, and vehicles) the information most relevant to the analysis would be linkages among entities. For this objective, information that does not provide linkages would be considered not relevant. Thus, in addition to critical thinking skills, the analyst needs to understand and be proficient in the application of specific analytical techniques such as link analysis or financial profiling.

6.1.3 Extract the essential message

Extracting the essential message is the ability to sort through the details of information and distinguish the essential from the non-essential. It also encompasses the ability to generate clear, concise statements that summarize the main point (the gist) of the information. The process is often automatic, because most people have extensive experience in attempting to get the main idea from what they read, see and hear. The automatic process usually works well if the amount of information is limited and the main points are stated clearly and unambiguously. However, critical thinking is needed when the information is extensive, is created in different formats and styles for different audiences, and the content has a high degree of complexity. The problem is further intensified when information is poorly presented with the main points not clearly discernable from the details.

The intelligence analyst typically deals with extensive amounts of information that is likely to be complex, is often ambiguous, may be prepared by someone from a different culture, and is not always presented clearly and simply. As a consequence, skill is required to extract the essential message from information and to summarize this message for future use in the analytical process. It is extremely useful to summarize a large amount of complex information with a simple statement so that the entire body of information need only be consulted subsequently to seek or verify specific details. Also, the gist serves as convenient shorthand to help communicate, is more easily remembered, and helps the analyst focus on the most important issues.

6.2 Organize information into premises

The skills associated with this function are: recognize patterns and relationships, and challenge assumptions. These skills are described in the paragraphs that follow.

6.2.1 Recognize patterns and relationships

An important function of intelligence analysis has been referred to in recent years as "connecting the dots" (Lahneman, 2006). While this expression is not very definitive, it does provide a general feeling for a skill that is important to the work of the analyst—recognizing and confirming patterns and relationships. A special aspect of this skill is establishing causes and effects that may be vital to understanding a situation, threat, process or set of events—who is sending suicide bombers into the crowded market places of the city, for example. This particular skill is one of recognizing patterns and relationships in the process of building premises that will lead, ultimately, to the development of hypotheses.

A critical task in the intelligence analysis process is the organization of information into premises—summarizing related items of information, results of data integration efforts, and/or information that answers a question into a summary statement that encompasses the central idea (premise) contained in the information. To complete this task successfully, the analyst must be able to recognize the patterns and relationships that serve as a logical basis for premise development.

6.2.2 Challenge assumptions

Information obtained for analysis may contain or be based on assumptions (ideas treated as facts but that are not yet supported by available evidence) that are not immediately obvious. On the other hand, the analyst might introduce, in the process of the analysis, assumptions that are mistakenly treated as evidence. Consequently, the analyst must have the capability to identify and challenge any and all assumptions, because they are very likely to be invalid or misleading.

The tendency to overlook or accept assumptions in an analysis might be related to biases introduced into the process, such as certain mind sets and expectations, but they can also be a function of simply not being attentive to their possible existence. The need to challenge assumptions arises mainly while organizing information into premises. Premises should be based on the evidence at hand, an effort that can be defeated by the inclusion of ideas and beliefs based on conjecture. Therefore, as a part of the premise formulation process, there should be a conscious effort to identify, challenge, and remove information that cannot be supported by the evidence at hand. This is an important analytical effort because the premises, once developed, provide the primary basis for hypothesis development.

6.3 Develop hypotheses

The skills associated with this function are: establish logical relationships; consider alternative perspectives; and counter biases, expectations, mind sets and oversimplification. These skills are described in the paragraphs that follow.

6.3.1 Establish logical relationships

The application of inductive logic to a set of premises to develop one or more hypotheses is at the heart of the intelligence analysis process. The hypothesis is a tentative explanation, subject to further testing, of a situation, process, threat, or activity of interest. Developing useful hypotheses requires skill in applying logical reasoning to a set of premises that have been developed from data organized and integrated for this purpose.

The critical aspect of this skill is that of organizing a set of premises into an argument that leads to an explanation that is based on the facts summarized in the premises, but that projects the explanation beyond these facts alone. That is, the analyst develops a hypothesis that fills in missing gaps to provide a more complete and more useful explanation. The set of hypotheses thus developed serve as the basis for guiding the collection of additional information to fill in the gaps with facts rather than conjecture. The establishment of logical relationships enables the intelligence analyst to link information to premises, premises to hypotheses, and hypotheses to inferences that can be acted on with confidence. The logical relationships are necessarily inductive in nature—going from the specifics to the general, permitting discovery of what was previously unknown. It is the tightness of this logic that provides the necessary discipline for the ultimate development of useful, valid inferences.

6.3.2 Consider alternative perspectives

This is the ability to develop explanations from different perspectives for the same information. An important component of this ability is to set aside one's own inclinations, values, beliefs, expectations, and preferences so as to develop explanations that cover the full range of possibilities. Some aspects of this skill have been called divergent thinking—generating different ideas about a topic from available information or knowledge. But while divergent thinking is characterized by spontaneous, free-flowing, unorganized idea generation, this skill requires the development of explanations from the deliberate consideration of a set of premises that have been systematically derived from available information.

Intelligence analysis relies on the development of alternative competing hypotheses. After a set of premises has been derived from information determined to be relevant and valid, alternative hypotheses are developed that define the full range of possible explanations for the information. This process requires the critical thinking skill of considering alternative perspectives. The resulting alternative hypotheses, then, serve to guide collection of the additional information needed to formulate a useful inference.

6.3.3 Counter biases, expectations, mind sets and oversimplification

Analysts are subject to the same biases, expectations, mindsets and oversimplifications that affect the thinking of all humans. While these negative influences might have limited impact on the lives that most of us live, they can be devastating to the work of the intelligence analyst. Consequently, analysts must develop the ability to understand and recognize the possible effects of these influences and to develop skills to keep them from distorting the products of analysis.

This skill involves the ability to continuously reevaluate one's view of the situation for these types of negative influences and to take the appropriate steps to eliminate them from the analysis. Although the types of influences addressed in this skill can enter the intelligence analysis process anywhere along the line, the primary concern is their role in hypothesis development and testing. Prior to this point, the tests for relevancy and validity should help assure the analyst that cognitive biases have had only a limited opportunity to enter the process. Now, as the analyst moves from strictly factual information to using conjecture in developing the most encompassing and useful hypotheses possible, these opportunities for distortion can operate most freely.

6.4 Test hypotheses

Testing hypotheses requires: considering value-cost-risk tradeoffs in seeking additional information, seeking disconfirming evidence, and assessing the strength of logical relationships. These skills are described in the paragraphs that follow.

6.4.1 Consider value-cost-risk tradeoffs in seeking additional information

A dilemma faced by intelligence analysts is whether to stop and report an inference based on available information, or to collect additional information. More information might produce an inference with greater usefulness at a higher level of confidence, but seeking additional information adds to intelligence costs and also risks a result that is not timely enough to be of value. This dilemma might be encountered early in the intelligence process or, more critically, later during the testing of hypotheses. This skill, then, is the ability to evaluate the need for new information by considering the value, cost and risk tradeoffs that are involved.

The analyst faces value-cost-risk tradeoffs principally during the stage of analysis in which hypotheses are being tested; this is a critical part of the process of developing a useful inference. Typically, one or more hypotheses would have been developed at this stage of the analysis and additional information might be required to help confirm or refute them. With limited time and resources available for collecting additional information, the analyst must employ these resources in the manner that will produce the greatest value for the resources expended. The analyst must also be sensitive to producing an inference in sufficient time and at a high enough level of confidence for it to be of use.

6.4.2 Seek disconfirming evidence

This skill is closely related to two skills addressed earlier—consider alternative perspectives and counter biases, expectations, mind sets, and oversimplification. Seeking disconfirming evidence is an important component of efforts taken to develop and test alternative competing hypotheses and is done in the face of biases that work to impede such efforts. A particularly important influence, confirmation bias, affects the development of alternative hypotheses by tending to prevent the analyst from seeking information other than what is likely to confirm a favored explanation.

The skill, then, is the ability to seek disconfirming evidence, particularly in the testing of hypotheses, when the more natural inclination is to seek confirming evidence. This skill is applied to intelligence analysis mainly during the testing of hypotheses. Assuming that the analysis has been performed effectively to this point, the analyst has two or more alternative explanations for the information at hand; testing these alternatives requires the collection of additional information that will ultimately result in selecting the most valid or producing some composite that is the most valid. To overcome our built-in human tendency to seek confirming evidence, the analyst needs to learn the techniques and discipline of seeking disconfirming evidence during the hypothesis testing process.

6.4.3 Assess the strength of logical relationships

The development of a hypothesis from a set of premises is based on the logical relationship that exists between premises and hypothesis. The relationship is necessarily one of inductive logic, in which the argument proceeds from the specifics (the premises) to the general (the hypothesis). The strength of the relationship depends on the extent of conjecture involved in making the jump from the facts as summarized in the premises and the hypothesis that goes beyond the premises to provide a more useful explanation. More conjecture leads to weaker relationships; less conjecture leads to stronger relationships. The most meaningful way to assess and convey the strength of this logical relationship is to provide a numerical probability estimate of the confidence one can have that the hypothesis or inference is true.

The critical thinking skill is that of assessing the strength of these relationships in a manner that provides a numerical probability of the validity of hypotheses and inferences. Critical thinking is required because the process is a subjective one—subjective conditional probability—calling for a careful and deliberate assessment. The process is necessarily subjective (and consequently requires critical thinking) because the analyst will hardly ever have the type of statistical evidence needed to provide a simple objective calculation of probability (one that does not require critical thinking). In applying subjective conditional probability, the analyst must answer the following question: Given this specific set of premises (the conditions), what is the probability that the hypothesis (or inference) is true?

As stated earlier in this paper, the objective of intelligence analysis is to develop inferences that can be acted on with confidence. For the product of intelligence analysis to be complete, therefore, it must produce an inference that provides the needed explanation and, also, an estimate of the level of confidence that the user can have in that inference. The goal is to provide the greatest level of detail at the highest level of confidence. However, this usually results in a tradeoff—greater detail typically comes at a lower level of confidence. Conversely, the analyst can provide a higher level of confidence but with less detail. Providing confidence assessments enables the analyst to best meet the needs of the user—more detail at lower confidence or less detail at higher confidence. To provide such estimates, the analyst must be capable of generating and communicating subjective conditional probability estimates.

7. Conclusions

In the last couple of decades a number of useful tools have been developed to support the intelligence process, encompassing the functions of data collection, evaluation, collation and integration. However, intelligence analysis remains highly dependent on the cognitive capabilities, specifically the critical thinking skills, of the human analyst. For this reason, it is important for the success of the process to understand the inherent capabilities and limitations of the analyst and, in particular, the challenges that must be overcome through the application of cognitive ergonomics to the design of analysis systems and in the training of critical thinking skills.

To better understand critical thinking and the efforts required to maximize its effectiveness, a model was developed that is sufficiently specific to enhance understanding and to permit empirical testing. The model identifies the role of critical thinking within the related fields of reasoning and judgment, which have been empirically studied since the 1950s and are better understood. It incorporates many ideas offered by leading thinkers in philosophy and education. It also embodies many of the variables discussed in the relevant literature (e.g., predisposing attitudes, experience, knowledge, and skills) and specifies the relationships among them. The model can, and has been, used to make testable predictions about the factors that influence critical thinking and about the associated psychological consequences. It also offers practical guidance to the development of training for critical thinking skills.

The model is based on the most recent versions of heuristic theory, the foundation of which is that two cognitive systems are used to make judgments. System 1, based on intuition, is a quick, automatic, implicit process that employs associational strengths to arrive at solutions automatically. System 2 is effortful, conscious, and deliberately controlled. The two systems run in parallel and work together, capitalizing on each other's strengths and compensating for their weaknesses. For example, one function of System 2, the controlled deliberate process, is to monitor the products of the automatic process, making adjustments to correct or block the judgment of System 1. If no intuitive response is accessible, System 2 will be the primary processing system used to arrive at a judgment.

Technology can now be employed extensively by intelligence analysts to extract meaning from available information, to support the performance of a variety of analyses, and to aid in the communication of analytical results to the users of intelligence. The design of future systems to support the intelligence process can benefit from cognitive ergonomics, specifically from what we now know about the nature and role of critical thinking. Moreover, findings about specific critical thinking skills can support the development of training systems and methods that best meet analyst performance requirements.

Research and experience to date in training and applying intelligence analysis skills suggest that the principal challenges that affect critical thinking are human limitations. Humans are limited in their capabilities to address complexity, by the biases they bring to the process, by their difficulties in handling uncertainty and, often, by the lack of relevant domain expertise. These limitations must be overcome by appropriately designed training systems and methods.

Recent research has identified the 11 critical thinking skills that are most important for successful intelligence analysis. They are presented below as they relate to the principal intelligence function they serve.

Assess and Integrate Information

- *Envision the end state of the analysis* and use that vision to guide and limit the analysis to those tasks most likely to attain the desired goal, checking on the process and products to ensure movement along the right path.
- *Assess and filter for relevance and validity,* examining information for its potential contribution to the objectives of the analysis.
- *Extract the essential message* by sorting through the details of information to distinguish the essential from the non-essential, and by generating clear, concise statements summarizing the main points.

Organize Information into Premises

- *Recognize patterns and relationships*, establishing causes and effects vital to understanding situations, threats, processes and events during the development of premises in an argument.
- *Challenge assumptions* so as to avoid ideas that might be treated as facts but that are not supported by available evidence or might be related to biases that have been introduced by mind sets or expectations.

Develop Hypotheses

- *Establish logical relationships* by applying inductive logic to derive one or more hypotheses from the set of premises summarizing facts derived from available information.
- *Consider alternative perspectives* by setting aside personal inclinations, values and expectations so as to develop explanations (hypotheses) that cover the full range of possibilities.
- *Counter biases, expectations, mind sets and oversimplification* by developing the ability to recognize the possible effects of these influences and developing techniques to keep them from distorting the products of analysis.

Test Hypotheses

- *Consider value-cost-risk tradeoffs in seeking additional information* to employ available resources in a manner that will produce the greatest value for the resources expended and the time available.
- *Seek disconfirming evidence* during the testing of hypotheses when the more natural inclination is to seek confirming evidence.
- *Assess the strength of logical relationships* in a manner that provides a numerical probability estimate of the confidence one can have in the validity of hypotheses and inferences.

8. References

American Philosophical Association (1990). *The Delphi report: Committee on pre-college philosophy* (ERIC Doc. No. ED 315 423.)

Auprasert, B. & Limpiyakorn, Y. (2008). Underlying cognitive complexity measure computation with combinatorial rules. *World Academy of Science, Engineering and Technology, 45*, 431-436.

Baddley A.D. (1986). *Working memory*. London/New York: Oxford University Press.

Baddley, A.D. (1996). Exploring the central executive. *Quarterly Journal of Experimental psychology, 49A*, 5-28.

Baron, J. B. & Sternberg, R. J. (1986). *Teaching thinking skills*. New York: W. H. Freeman and Company.

Engle, R.W. & Kane, M. J. (2004). Executive attention, working memory capacity, and a two-factor theory of cognitive control. In B. Ross (Ed.). *The psychology of learning and motivation, 44*, 145-199. NY: Elsevier.

Fingar, T. (2006). US Intelligence reform and the analytic community: challenges and opportunities. Proceedings of the IALEIA World Intelligence Training Conference, Mexico City, Mexico, April 24-28.

Fischer, S.C. & Spiker, V.A. (2000). *Application of a theory of critical thinking to Army command and control*. Alexandria, VA: US Army Research for the Behavioral and Social Sciences.

Fischer, S.C. & Spiker, V.A. (2004). *Critical thinking training for Army officers. Vol II: A model of critical thinking* (Phase II Final Report). Alexandria, VA: US Army Research for the Behavioral and Social Sciences.

Fischer, S.C., Spiker, V.A. and Riedel, S.L (2009). *Critical thinking training for Army officers. Vol III: Development and assessment of a web-based training program.* (Research Report 1883). Alexandria, VA: US Army Research for the Behavioral and Social Sciences.

Halpern, D. F. (1996). *Thought and knowledge: An introduction to critical thinking*. Mahwah, New Jersey: Lawrence Erlbaum.

Harris, D.H. (2006a). Overcoming the cognitive challenges of intelligence analysis. *Proceedings of the IALEIA World Intelligence Training Conference,* International Association of Law Enforcement Intelligence Analysts, Mexico City, April 24-28.

Harris, D.H. (2006b). Critical thinking strategies for intelligence analysis. Invited Workshop Presentation to the United Nations Regional Criminal Intelligence Conference, Belgrade, Serbia, December 12-13.

Harris, D.H. (2011). Critical thinking training for intelligence analysis. *Journal of the International Association of Law Enforcement Intelligence Analysts, 20 (1), 76-90.*

Heuer, R. (1999). *Psychology of intelligence analysis.* Washington, DC: Center for the Study of Intelligence, available online at http://www.cia.gov/csi/books/19104/art1.html.

Kahneman, D. (2003). A perspective on judgment and choice: Mapping bounded rationality. *American Psychologist, 58*, 697-720.

Klein, G. (1999). *Sources of power*. Cambridge, MA: MIT Press.

Klein, G. (2011). Critical thoughts about critical thinking. *Theoretical Issues in Ergonomics Science,* vol. 12 (3), 210-224.

Klein, G.A., Moon, B., & Hoffman, R.R. (2006). Making sense of sensemaking 2: A macrocognitive model. *IEEE Intelligent Systems, 21* (5), 88-92.

Lahneman, W.J. (2006). *The future of intelligence analysis: Volume I final report.* School of Public Policy, University of Maryland.

Meehl, P. E. (1954). *Clinical versus statistical prediction: A theoretical analysis and a review of the evidence*. Minneapolis: University of Minneapolis Press.

Miller, G. A. (1956). The magical number seven plus or minus two. *The Psychological Review,* vol. 63, pp. 81-97

Moore, D. T. (2007). Critical thinking and intelligence analysis. Center for Strategic Intelligence Research, National Defense Intelligence College: Washington, DC.

Packer, G. (2006). Knowing the enemy. *The New Yorker,* 60-69, December 18.

Paul, R. & Elder, L. (2001). *Critical thinking – Tools for taking charge of your learning and your life*. New Jersey: Prentice Hall.

Simon, H. (1957). *Models of man: Social and rational.* New York: Wiley.

Thaler, R.H. & Sunstein, C.R. (2008). *Nudge: Improving decisions about health, wealth, and happiness*. New York: Penguin Press.

Watson, G. & Glaser, E. M. (1980). *Critical thinking appraisal.* New York: Psychological Corporation.